Less is More

Jason Hickel is an economic anthropologist, Fulbright Scholar, and Fellow of the Royal Society of Arts. He is originally from Eswatini (Swaziland) and spent a number of years with migrant workers in South Africa, writing about exploitation and political resistance in the wake of apartheid. He has authored three books, including most recently *The Divide: A Brief Guide to Global Inequality and its Solutions*. He writes regularly for the *Guardian*, Al Jazeera and *Foreign Policy*, serves as an advisor for the Green New Deal for Europe and sits on the *Lancet* Commission for Reparations and Redistributive Justice. He lives in London.

ALSO BY JASON HICKEL

The Divide:
A Brief Guide to Global Inequality and its Solutions

Less is More

How Degrowth Will
Save the World

JASON HICKEL

PENGUIN BOOKS

PENGUIN BOOKS

UK | USA | Canada | Ireland | Australia
India | New Zealand | South Africa

Penguin Books is part of the Penguin Random House group of companies
whose addresses can be found at global.penguinrandomhouse.com

First published by William Heinemann in 2020
First published in paperback by Windmill Books in 2021
Published in Penguin Books 2022
011

Typeset by Jouve (UK), Milton Keynes
Printed and bound in Great Britain by Clays Ltd, Elcograf S.p.A.

The authorised representative in the EEA is Penguin Random House Ireland,
Morrison Chambers, 32 Nassau Street, Dublin D02 YH68

A CIP catalogue record for this book is available from the British Library

ISBN: 978–1–786–09121–5

www.greenpenguin.co.uk

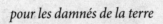

pour les damnés de la terre

We don't have a right to ask whether we are going to succeed or not. The only question we have a right to ask is what's the right thing to do? What does this Earth require of us if we want to continue to live on it?

–Wendell Berry

Table of Contents

Introduction

Welcome to the Anthropocene

My heart is moved by all I cannot save. So much has been destroyed. I have cast my lot with those who, age after age, perversely, with no extraordinary power, reconstitute the world.

Adrienne Rich

Sometimes these realisations sneak up on you, like a quiet memory – just the slightest hint that something isn't right.

When I was growing up in Eswatini, the small country in southern Africa formerly known as Swaziland, my family had a rickety old Toyota pickup – the kind that was ubiquitous in the region in the 1980s. After long drives it was my job to help clear the front grille of all the insects that accumulated there. Sometimes they were piled three deep: butterflies, moths, wasps, grasshoppers, beetles of every conceivable size and colour – dozens if not hundreds of species. I remember my dad telling me that the insects on Earth weighed more than all the other animals put together, including humans. I marvelled at this idea, and found it somehow heartening. As a child I worried about the fate of the living world, as I think many children do – so this story about the insects made me feel that everything was going to be OK. It was comforting to be reminded of the seemingly inexhaustible abundance of life. This fact would drift to mind on hot nights while we sat outside on the porch of our little tin-roofed house, hoping for a breeze, watching moths and beetles swarm around the light, dodging the bats that would sometimes swoop through to snatch a meal. I became fascinated with insects. At one point I tried to identify all the different species around our home, running about with pen and little notebook in hand. In the end I had to give up. There were too many to count.

My dad still shares that old story about the insects from time to time – always in an excited tone, in the way that dads do, like it's a new fact he's just discovered. But these days it doesn't quite ring true. Things feel different, somehow. When I've returned to southern Africa for research in recent years, the car turns out more or less clean even after long journeys. Maybe a few flies here and there, but nothing at all like before. Perhaps it's just that the insects loom large in my childhood memories. Or perhaps there's something more troubling afoot.

*

In late 2017, a team of scientists reported some strange and rather alarming findings. They had been meticulously counting insect numbers in German nature reserves for decades. This is something that very few scientists had taken the time to do – the sheer abundance of insects makes such an exercise seem unnecessary – so everyone was curious to see what would come of it. The results were devastating. The team found that three-quarters of flying insects in Germany's nature reserves had vanished over the course of twenty-five years – due, they concluded, to the conversion of surrounding forests to farmland, followed by the intensive use of agricultural chemicals.

The study went viral, capturing headlines around the world. 'We appear to be making vast tracts of land inhospitable to most forms of life, and are currently on course for ecological Armageddon,' one of the scientists said. 'If we lose the insects then everything is going to collapse.'[1] Insects are essential to pollination and plant reproduction, they break down organic waste and turn it into soil, and they provide a vital source of food for thousands of other species. As insignificant as they may seem, they are key nodes in the web of life. As if to confirm these fears, a few months later two studies reported that falling insect populations had contributed to a dramatic decline of birds on farmland in France. Average numbers had fallen by a third in only fifteen years, with some species collapsing by much more: meadow pipits declined 70% over this period, while partridges declined 80% over twenty-three years.[2] In the same year, news out of China reported that insect die-offs had triggered a pollination crisis. Eerie photographs emerged of workers going from plant to plant, pollinating crops by hand.

The problem isn't unique to these regions. Insect decline appears to be widespread. It is difficult to assess trends on continental or global scales, but the evidence does not look good. Researchers have found that the abundance of terrestrial insects has been declining by around 9% per decade,[3] and at least one in every ten species is now at risk of extinction.[4] These are alarming figures. And it raises concerns about the possibility of 'cascading extinctions', whereby the destruction of one species may trigger the decline of others, exacerbating biodiversity loss in unpredictable ways.[5] The crisis has become so severe that in 2020, scientists published a 'warning to humanity' about the fate of insects. 'With insect extinctions, we lose much more than species', they wrote. We lose 'large parts of the tree of life', and such losses 'lead to the decline of key ecosystem services on which humanity depends.'[6] Echoing these sentiments, a recent symposium of world experts on insect biodiversity produced a report that opened with four simple but ominous words: 'Nature is under siege'.[7]

*

This is not a book about doom. It is a book about hope. It's about how we can shift from an economy that's organised around domination and extraction to one that's rooted in reciprocity with the living world. But before we begin that journey, it's important that we grasp what's at stake. The ecological crisis happening around us is much more serious than we generally assume. It's not just one or two discrete issues, something that could be solved with a targeted intervention here and there while everything else carries on as normal. What's happening is the breakdown of multiple, interconnected systems – systems on which human beings are fundamentally dependent. If you're already familiar with what's going on, you may want to skim over this part. If not, brace yourself. It's not just the insects.

Living in an age of mass extinction

Perhaps it seemed like a good idea at the time: transfer land to big companies, rip up any hedges and trees and plant it all with a single crop, spray it from aeroplanes and harvest with giant combines. Beginning in the middle of the twentieth century, whole landscapes were remade according to the totalitarian logic of industrial profit, most of it for livestock feed, with the goal of maximising extraction. They called it the Green Revolution but, from the perspective of ecology, there was nothing 'green' about it. By reducing complex ecological systems to a single dimension, everything else became invisible. Nobody noticed what was happening to the insects and the birds. Or even to the soil itself.

If you've ever picked up a handful of rich, dark, fragrant soil, you'll know that it's crawling with life: worms, grubs, insects, fungus and millions of microorganisms. That life is what makes soils resilient and fertile. But over the past half-century, industrial agriculture, with its reliance on aggressive ploughing and chemical inputs, has been killing soil ecosystems at a rapid clip. Agricultural soil under industrial tillage is eroding more than 100 times faster than it is being formed.[8] In 2018, a scientist from Japan made the effort to sort through evidence on earthworm populations from around the world. He found that on industrial farms earthworm biomass had plunged by a dramatic 83%. And as the earthworms died off, the organic content of soils collapsed by more than half. Our soils are being turned into lifeless dirt.[9]

Something similar is happening in our oceans. When we go to the supermarket, we take for granted that we'll find all the seafood we love: cod, hake, haddock, salmon, tuna – species that are central to human diets all around the world. But this easy certainty is beginning to crumble. Recent figures show that 34% of fish stocks in the world's marine fisheries are now overfished and in decline – three times more than in the 1970s.[10] In the UK, haddock have fallen to 1% of their nineteenth-century volume; halibut, those magnificent giants of the sea, to one-fifth of 1%.[11] Fish catches are beginning to decline around the world, for the first time in recorded history.[12] In the Asia-Pacific, exploitable fish stocks could decline to zero by 2048 if current trends continue.[13]

Most of this is due to aggressive overfishing: just as with agriculture, corporations have turned fishing into an act of warfare, using industrial megatrawlers to scrape the seafloor in their hunt for increasingly scarce fish, hauling up hundreds of species in order to catch the few that have 'market value', turning coral gardens and colourful ecosystems into lifeless plains in the process. Whole ocean landscapes have been decimated in the scramble for profit. But there are also other forces at work. Farming chemicals like nitrogen and phosphorous are flowing into rivers and ending up in the sea, creating giant algae blooms that cut off oxygen to the ecosystems that lie beneath them. Vast 'dead zones' sprawl along the coastlines of industrialised regions like Europe and the United States. Once churning with life, many of our seas are becoming eerily empty, populated more by plastic than by fish.

Oceans are also being affected by climate change. More than 90% of the heat from global warming gets absorbed into the sea.[14] Oceans act like a buffer, protecting us from the worst effects of our emissions. But they are suffering as a result: as oceans heat up, nutrient cycles are being disrupted, food chains broken, and many marine habitats are dying off.[15] At the same time, carbon emissions are causing oceans to become more acidic. This is a problem, because ocean acidification has driven mass extinction events a number of times in the past. It played a major role in the last extinction event, 66 million years ago, when ocean pH dropped by 0.25. That small shift was enough to wipe out 75% of marine species. On our present emissions trajectory, ocean pH will drop by 0.4 by the end of the century.[16] We *know* what problems this could cause. We can see it coming. In fact, it's already beginning to play out in real time: marine animals are disappearing at twice the rate that land animals are.[17] Many coral ecosystems are being bleached into dead, colourless skeletons.[18] Divers have reported that even remote reefs once teeming with life are now plagued by the stench of decomposing flesh.

*

What begins as a vague inkling about moths and beetles, the flickers of a childhood memory, turns into a crippling realisation, like a blow to the gut. We are sleepwalking into a mass extinction event – the sixth in our planet's history, and the first to be caused by human economic activity. Extinctions are now occurring 1,000 times faster than the normal background rate.[19]

A few years ago, virtually no one was talking about this. Like my dad with his insect stories, everyone just assumed that the web of life would always be intact. Now the situation is so severe that the United Nations has set up a special task force to monitor it: the Intergovernmental Science-Policy Platform on Biodiversity and Ecosystem Services (IPBES). In 2019 it published its first comprehensive report – a groundbreaking assessment of the planet's living species, drawing on 15,000 studies from around the world and representing the consensus of hundreds of scientists. It found an accelerating rate of global biodiversity decline, unprecedented in human history. Around one million species are now at risk of extinction, many within decades.[20]

I keep staring at these numbers, but I can't get them to make any sense. It all feels so surreal, like a fever dream where the world seems strange, unfamiliar and out of proportion. Robert Watson, the Chair of the IPBES, called the UN report 'ominous'. 'The health of ecosystems on which we and all other species depend is deteriorating more rapidly than ever,' he said. 'We are eroding the very foundations of our economies, livelihoods, food security, health and quality of life worldwide.' Anne Larigauderie, the IPBES executive secretary, put it even more bluntly: 'We are currently, in a systematic manner, exterminating all non-human living beings.' Scientists are not known for using strong language. They prefer to write in a neutral, objective tone. But reading through these reports, one can't help noticing that many of them have felt compelled to shift registers. A recent study published in the prestigious *Proceedings of the National Academy of Sciences* – a serious, stuffy journal – described the extinction crisis as 'biological annihilation', and concluded that it represents a 'frightening assault on the foundations of human civilisation'. 'Humanity will eventually pay a very high price,' the authors wrote, 'for the decimation of the only assemblage of life that we know of in the universe.'[21]

*

This is the thing about ecology: everything is interconnected. It's difficult for us to grasp how this works, because we're used to thinking of the world in terms of individual parts rather than complex wholes. In fact, that's even how we've been taught to think of ourselves – as individuals. We've forgotten how to pay attention to the relationships between things. Insects necessary for pollination; birds that control crop pests, grubs and worms essential to soil fertility; mangroves that purify water; the corals on which fish populations depend: these living systems are not 'out there', disconnected from humanity. On the contrary: our fates are intertwined. They are, in a real sense, *us*.

It is impossible to adequately understand our ecological crisis with the same reductive thinking that caused it in the first place. This is particularly clear when it comes to climate change. We tend to think about climate change as primarily a matter of temperature. Many people are not particularly concerned about this, because our everyday experience with temperature is that a few degrees doesn't really make that much of a difference. But temperature is just the beginning – it's the loose thread on the sweater.

Some of the consequences of temperature rise are obvious, since we can see and experience them directly. The number of extreme storms that happen each year has doubled since the 1980s.[22] They now hit so frequently that even extraordinary spectacles blur together in our memories. If you remember, 2017 alone clobbered the Americas with some of the most destructive hurricanes on record. Harvey laid waste to huge swathes of Texas; Irma left Barbuda virtually uninhabitable; Maria plunged Puerto Rico into months of darkness, and wiped out 80% of the island's crop value. These were Category 5 hurricanes – the most severe type. Storms like these should happen only once in a generation. But in 2017 they rolled in one after another, leaving mayhem and destruction in their wake.

Rising temperatures have also triggered deadly heatwaves. The heatwave that struck Europe in 2003 killed a staggering 70,000 people in just a few days. France was hit hardest, with temperatures soaring over 40°C for more than a week. Wheat crops declined by 10% as drought ravaged the continent. Moldova saw its whole harvest decimated. Three years later it happened again, breaking temperature records across northern Europe. In 2015, heatwaves in India and Pakistan sustained temperatures over 45°C and killed more than 5,000 people. In 2017, a heatwave across Portugal triggered wildfires that ripped through the country's forests. Roads became graveyards as people roasted to death in their cars while trying to flee. Smoke blackened the skies as far away as London. In 2020, bush fires in Australia forced people to take refuge on beaches, in scenes reminiscent of an apocalyptic film. As many as one billion wild animals were killed. Horrific images emerged of landscapes strewn with charred kangaroos and koalas.

Events like these feel real and tangible. They become media headlines. But the more dangerous aspects of climate change do not. At least not yet. So far we've only barely breached 1°C over pre-industrial levels. On our current trajectory, as of 2020, we are on track to reach a rise of up to 4°C by the end of the century. If we factor in countries' pledges to cut emissions under the Paris Agreement – which are voluntary and non-binding – global temperatures will still rise by as much as 3.3°C. These are not incremental changes. Humans have never lived on such a planet. That deadly heatwave that struck Europe in 2003? That will be a normal summer. Spain, Italy and Greece will turn into deserts, with climates more like the Sahara than the Mediterranean as we know it. The Middle East will be cast into permanent drought.

At the same time, rising seas will change our world almost beyond recognition. So far, sea levels are up about 20cm since 1900. Even this apparently small rise has made flooding more frequent and storm surges more dangerous. When Hurricane Michael smashed into the United States in 2018, it brought a 14-foot surge that turned parts of the Florida coastline into a hellscape of shattered houses and twisted metal. If we carry on with business as usual, all of this will get much worse. In fact, even if we meet the Paris goal of keeping temperature rises to no more than 2°C, sea levels are projected to go up another 30 to 90cm by the end of the century.[23] Given the damage that 20cm has caused, it's difficult to imagine what things will be like when it's up to four times higher than it is right now. The storm surges alone will be catastrophic. The wall of waves unleashed by Hurricane Michael will seem quaint by comparison. And if temperatures rise by 3°C or 4°C, sea levels will go up by as much as 100cm, and possibly 200cm. Many of the planet's coastal regions will be underwater. Much of Bangladesh, home to 164 million people, will disappear. Without massive seawall infrastructure, cities like New York and Amsterdam will be permanently flooded, as will Jakarta, Miami, Rio and Osaka. Under these conditions, countless people would be forced to flee their homes. All this century.

And yet, as disastrous as all of this is likely to be, perhaps the most concerning impact of climate change has to do with something much more quotidian: food. Half of Asia's population depends on water that flows from Himalayan glaciers – not only for drinking and other household needs but also for agriculture. For thousands of years, the run-off from those glaciers has been replenished each year by new ice. But now the ice is melting faster than it is being replaced. If we hit 3°C or 4°C of warming, most of those glaciers will be gone before the end of the century, ripping the heart out of the region's food system and leaving 800 million people in trouble. In southern Europe, Iraq, Syria and much of the rest of the Middle East, extreme droughts and desertification could render whole regions inhospitable to agriculture. Major food-growing regions in the US and China will also take a hit. According to NASA, droughts in the American plains and in the South-west could turn these regions into dust bowls.[24]

The impact of temperature rise on crops varies by species and by region, but on average, yields of major crops like wheat, rice, maize and soybean are likely to decline by 3-7% per degree C.[25] This could cause major problems, particularly in tropical regions. Under normal circumstances, regional food shortages can be covered by surpluses from elsewhere on the planet. But climate breakdown could trigger shortages on multiple continents at once. According to the Intergovernmental Panel on Climate Change (IPCC), warming more than 2 degrees is likely to cause 'sustained food supply disruptions globally'. As one of the lead authors of the report put it: 'The potential risk of multi-breadbasket failure is increasing.' Add this to soil depletion, pollinator die-off and fishery collapse, and we're looking at spiralling food emergencies.

This will have serious implications for global political stability. Regions affected by food shortages will see mass displacement as people migrate in search of stable food supplies. In fact, it's happening already.[26] Many of those fleeing places like Guatemala and Somalia are doing so because their farms are no longer viable. The international system is already straining, with 65 million people displaced from their homes by wars and droughts – more than at any time since the Second World War. As migration pressures build, politics are becoming more polarised, fascist movements are on the march, and international alliances are beginning to fray. Factor in escalating displacement due to famines, storms and rising seas, plus dwindling arable farmland, and there's no predicting what conflagrations might occur.

*

Ecosystems are complex networks. They can be remarkably resilient under stress, but when certain key nodes begin to fail, knock-on effects reverberate through the web of life. This is how mass extinction events unfolded in the past. It's not the external shock that does it – the meteor or the volcano: it's the cascade of internal failures that follows. It can be difficult to predict how this kind of thing plays out. Things like tipping points and feedback loops make everything much riskier than it otherwise might be. This is what makes climate breakdown so concerning.

Take the polar ice caps, for example. Ice functions like a giant reflector, bouncing light from the sun back out into space. This is known as the albedo effect. But as ice sheets disappear and reveal the darker landscapes and oceans beneath, all that solar energy gets absorbed and radiated as heat into the atmosphere. This drives yet further warming, which causes the ice to melt even faster – completely irrespective of human emissions. In the 1980s, Arctic sea ice covered an average of about 7 million square kilometres. As I write this it's down to about 4 million.

Feedback loops affect forests, too. As the planet heats up, forests become drier and more vulnerable to fire. When forests burn they release carbon into the atmosphere, and we lose them as a sink for future emissions. This exacerbates global warming, but it also has a direct impact on rainfall. Forests literally *produce* rain. The Amazon, for instance, exhales some 20 billion tons of water vapour into the atmosphere every day, like an enormous river flowing invisibly into the sky. Much of it ends up raining back down onto the forest, but it also produces rain much further afield – across South America and even as far north as Canada. Forests are critical to our planet's circulatory system; they are like giant hearts that pump life-giving water around the world.[27] As forests die off, droughts become more common, and forests in turn become yet more vulnerable to fire. The speed at which this is happening is frightening.

13

On our current trajectory, most rainforests will wither away into savannah before the end of this century.

In some cases, tipping points work so rapidly that whole systems can collapse in a very short period of time. Scientists worry in particular about a phenomenon known as marine ice-cliff instability. In the past, most climate models have assumed that even if global warming locks in the total melting of the West Antarctic ice sheet, the process of disintegration will stretch out over a couple of centuries. But in 2016, two American scientists – Rob DeConto and David Pollard – published an article in the journal *Nature* pointing out that it may well happen a lot faster. Ice sheets are thicker in the middle than they are around the edges, so as icebergs break off they expose taller and taller ice cliffs. This is a problem, because taller ice cliffs can't support their own weight: once they're exposed they begin to buckle, one after the other, in a domino effect, like skyscrapers collapsing. This dynamic could cause ice sheets to disintegrate faster than previously assumed.[28]

If this plays out, the West Antarctic ice sheet alone could add another metre or more to sea-level rise, in our lifetime. If the same thing happens to Greenland, it would be worse still. Coastal cities could be flooded at a pace that would make adaptation very difficult. Large parts of Kolkata, Shanghai, Mumbai and London would be swamped, along with much of the world's economic infrastructure. This would be a catastrophe of almost unimaginable scale. And we know it can happen, because it's happened before. It occurred at the end of the last ice age, in fact. Scientists who study ice-cliff dynamics have been loudly critical of governments for not accounting for this risk in their climate models.

All of this complexity opens up real questions about our ability to control global temperatures. Some scientists worry we may not be able to 'park' temperature increases at 2 degrees, as the Paris Agreement assumes. If we heat to 2 degrees, we might trigger cascades that could spiral out of control and push the Earth into a permanent 'hothouse state'. Temperatures could soar far above the target threshold, and we would be powerless to stop it.[29] In light of these risks, the only rational response is to do everything possible to keep warming to no more than 1.5°C. And that means cutting global emissions to zero much, much faster than anyone is presently planning.

Behind the eco-fact

This isn't the first time you've heard all of this, of course. If you're reading this book, it's probably because you're already concerned. You've already read dozens of stomach-churning facts about the crisis we're in. You know something is terribly wrong. I don't need to convince you. That's not what this book is for.

The philosopher Timothy Morton has likened our obsession with eco-facts to the nightmares suffered by people with post-traumatic stress disorder, or PTSD. In PTSD dreams, you relive your trauma and wake up viscerally terrified, sweating and shaking. For some reason the nightmares happen over and over again. Sigmund Freud argued that this is your mind's attempt to ameliorate your fear by trying to insert you into the moments right before the trauma happened. The idea is that if you're able to anticipate the traumatic event, you might be able to avoid it – or at least prepare yourself psychologically. Morton thinks our eco-facts serve a similar function. By endlessly repeating terrifying eco-facts, on some subconscious level we're trying to insert ourselves into a fictional moment right before collapse happens, so we can see it coming and do something about it. At least we'll feel prepared when it arrives.[30]

In this sense, eco-facts carry a double message. On the one hand they cry out, urging us to wake up and act right now. But at the same time they imply that the trauma is not yet fully here – that there's still time to avert disaster. This is what makes them so beguiling, so reassuring, and why we seem strangely to crave more of them. The danger of this is that we will all be lulled into waiting around for the facts to become more extreme. Once we reach that point, we tell ourselves, we'll finally get around to doing something about it. But the ultimate eco-fact is never going to arrive. It's never going to be good enough. Just as in the PTSD dream, eco-facts never work as they're supposed to. They always fail, and in the end we wake up crying in the middle of the night, shivering with unspeakable fear, because on some deep level we know that the trauma has already arrived. We're already in the middle of it. We are living in a world that is dying.

The facts have been piling up for decades. They become more elaborate, and more concerning, with each passing year. And yet for some reason we have been unable to change course. The past half-century is littered with milestones of inaction. A scientific consensus on anthropogenic climate change first began to form in the mid-1970s. The first international climate summit was held in 1979, three years before I was born. The NASA climate scientist James Hansen gave his landmark testimony to the US Congress in 1988, explaining how the combustion of fossil fuels was driving climate breakdown. The UN Framework Convention on Climate Change (UNFCCC) was adopted in 1992 to set non-binding limits on greenhouse gas emissions. International climate summits – the UN Congress of Parties – have been held annually since 1995 to negotiate plans for emissions reductions. The UN framework has been extended three times, with the Kyoto Protocol in 1997, the Copenhagen Accord in 2009, and the Paris Agreement in 2015. And yet global CO_2 emissions continue to rise year after year, while ecosystems unravel at a deadly pace.

Even though we have known for nearly half a century that human civilisation itself is at stake, there has been no progress in arresting ecological breakdown. None. It is an extraordinary paradox. Future generations will look back on us and marvel at how we could have known exactly what was going on, in excruciating detail, and yet failed to solve the problem.

What explains this inertia? Some will point to fossil fuel companies and the vice-like grip they have on our political systems. And certainly there is truth to this. Some of the larger companies, despite knowing about the dangers of climate breakdown long before it was part of the public debate, have bankrolled politicians who have either denied the science outright or sought to obstruct meaningful action whenever possible. It is in large part thanks to their efforts that the international climate treaties are not legally binding, for they have lobbied vigorously against such a move. And they have waged an extraordinarily successful disinformation campaign that for decades eroded public support for climate action, particularly in the United States, the one country that could feasibly lead a global transition.

Fossil fuel companies, and the politicians they have bought, bear significant responsibility for our predicament. But this alone doesn't explain our failure to act. There's something else – something deeper. Our addiction to fossil fuels, and the antics of the fossil fuel industry, is really just a symptom of a prior problem. What's ultimately at stake is the economic system that has come to dominate more or less the entire planet over the past few centuries: capitalism.

*

Mention the word capitalism, and people immediately get their hackles up. Everyone has strong feelings about it, one way or the other, often for good reasons. But whatever we might think of capitalism, it's important to have a clear-eyed view of what it is and how it works.

We have a tendency to describe capitalism with familiar, well-worn words like 'markets' and 'trade'. But this isn't quite accurate. Markets and trade were around for thousands of years before capitalism, and they are innocent enough on their own. What makes capitalism different from most other economic systems in history is that it's organised around the imperative of constant expansion, or 'growth': ever-increasing levels of industrial production and consumption, which we have come to measure in terms of Gross Domestic Product (GDP).[31] Growth is the prime directive of capital. And as far as capital is concerned, the purpose of increasing production is not primarily to meet specific human needs, or to improve social outcomes. Rather, the purpose is to extract and accumulate an ever-rising quantity of profit. That is the overriding objective. Within this system, growth has a kind of totalitarian logic to it: every industry, every sector, every national economy must grow, all the time, with no identifiable end-point.

It can be difficult to grasp the implications of this. We tend to take the idea of growth for granted because it sounds so *natural*. And it is. All living organisms grow. But in nature there is a self-limiting logic to growth: organisms grow to a point of maturity, and then maintain a state of healthy equilibrium. When growth fails to stop – when cells keep replicating just for the sake of it – it's because of a coding error, like what happens with cancer. This kind of growth quickly becomes deadly.

Under capitalism, global GDP needs to keep growing by at least 2% or 3% per year, which is the minimum necessary for large firms to maintain rising aggregate profits.[32] That might seem like a small increment, but keep in mind that this is an exponential curve, and exponential curves have a way of sneaking up on us with astonishing speed. Three per cent growth means doubling the size of the global economy every twenty-three years, and then doubling it again from its already doubled state, and then again, and again. This might be OK if GDP were just plucked out of thin air. But it's not. It is coupled to energy and resource use, and has been for the entire history of capitalism. There's a bit of give between the two, but not much. As production increases, the global economy churns through more energy, resources and waste each year, to the point where it is now dramatically over-shooting what scientists have defined as safe planetary boundaries, with devastating consequences for the living world.[33]

But, contrary to what the language of the Anthropocene implies, the ecological crisis is not being caused by all human beings equally. This is a crucial point to grasp. As we will see in Chapter 2, low-income countries, and indeed most countries in the global South, remain well within their fair share of planetary boundaries. In fact, in many cases they need to *increase* energy and resource use in order to meet human needs. It's high-income countries that are the problem here, where growth has become completely unhinged from any concept of need, and has long been vastly in excess of what is required for human flourishing. Global ecological breakdown is being driven almost entirely by excess growth in high-income countries, and in particular by excess accumulation among the very rich, while the consequences hurt the global South, and the poor, disproportionately.[34] Ultimately, this is a crisis of inequality as much as anything else.

*

We know exactly what we need to do in order to avert climate breakdown. We need to actively scale down fossil fuels and mobilise a rapid rollout of renewable energy – a global Green New Deal – to cut world emissions in half within a decade and get to zero before 2050. Keep in mind that this is a global average target. High-income nations, given their greater responsibility for historical emissions, need to do it much more quickly, reaching zero by 2030.[35] It is impossible to overstate how dramatic this is; it is the single most challenging task that humanity has ever faced. The good news is that it is absolutely possible to achieve. But there's a problem: scientists indicate that it cannot be done quickly enough to keep temperatures under 1.5°C, or even 2°C, if high-income economies continue to pursue growth at the same time.[36] Why? Because more growth means more energy demand, and more energy demand makes it all the more difficult – impossible, in fact – to roll out enough renewables to cover it in the short time we have left.[37]

Even if this wasn't a problem, we must ask ourselves: once we have 100% clean energy, what are we going to do with it? Unless we change how our economy works, we'll keep doing exactly what we are doing with fossil fuels: we'll use it to power continued extraction and production, at an ever-increasing rate, placing ever-increasing pressure on the living world, because that's what capitalism requires. Clean energy might help deal with emissions, but it does nothing to reverse deforestation, overfishing, soil depletion and mass extinction. A growth-obsessed economy powered by clean energy will still tip us into ecological disaster.

The tricky part is that it seems we have little choice about this. Capitalism is fundamentally *dependent* on growth. If the economy doesn't grow it collapses into recession: debts pile up, people lose their jobs and homes, lives shatter. Governments have to scramble to keep industrial activity growing in a perpetual bid to stave off crisis. So we're trapped. Growth is a structural imperative – an iron law. And it has ironclad ideological support: politicians on the left and right may bicker about how to distribute the yields of growth, but when it comes to the pursuit of growth itself they are united. There is no daylight between them. Growthism, as we might call it, stands as one of the most hegemonic ideologies in modern history. Nobody stops to question it.

It is because of their commitment to growthism that our politicians find themselves unable to take meaningful action to stop ecological breakdown. We have dozens of ideas for how to fix the problem, but we dare not implement them because doing so might undermine growth. And in a growth-dependent economy, that cannot be allowed to happen. Instead, the very newspapers that carry harrowing stories about ecological breakdown also report excitedly on how GDP is growing every quarter, and the very politicians who wring their hands about climate breakdown also call dutifully for more industrial growth every year. The cognitive dissonance is striking.

Some people try to reconcile this tension by leaning on the hope that technology will save us – that innovation will make growth 'green'. Efficiency improvements will enable us to 'decouple' GDP from ecological impact so we can continue growing the global economy for ever without having to change anything about capitalism. And if this doesn't work, we can always rely on giant geo-engineering schemes to rescue us in a pinch.

It's a comforting fantasy. In fact, I once believed it myself. But when I began to peel back the layers of nice-sounding rhetoric, I realised that it is just that – a fantasy. I have been researching this for a number of years, in collaboration with colleagues in ecological economics. In 2019 we published a review of existing evidence; and in 2020 scientists ran a number of meta-analyses, examining data from hundreds of studies.[38] I'll explain the details in Chapter 3, but the conclusions boil down to this: 'green growth' is not a thing. It has no empirical support. These findings were an epiphany for me, and forced me to change my position. In an era of ecological emergency, we cannot afford to build policy around fantasies.

Don't get me wrong. Technology is absolutely essential in the fight against ecological breakdown. We need all the efficiency improvements we can get. But scientists are clear that they will not be enough, on their own, to fix the problem. Why? Because in a growth-oriented economy, efficiency improvements that *could* help us reduce our impact are harnessed instead to advance the objectives of growth – to pull ever-larger swathes of nature into circuits of extraction and production. It's not our technology that's the problem. It's growth.

Stirrings

Fredric Jameson once famously said that it is easier to imagine the end of the world than to imagine the end of capitalism. This isn't so surprising, really. After all, capitalism is all we know. Even if we were to somehow put an end to it, what would happen afterwards? What would we replace it with? What would we do on the day after the revolution? What would we call it? Our capacity for thought – and even our language – stops at the boundaries of capitalism, and beyond lies a terrifying abyss.

How odd. We are a culture that is enamoured of newness, obsessed with invention and innovation. We claim to celebrate creative, out-of-the-box thinking. Certainly we would never say of a smartphone or a piece of art, 'This is the best gadget or painting that has ever been created and it will never be surpassed, and we shouldn't even try!' It would be naïve to underestimate the power of human creativity. So why is it that, when it comes to our economic system, we have so readily swallowed the line that capitalism is the only possible option and we shouldn't even *think* about creating something better? Why are we so wedded to the dusty dogmas of this old sixteenth-century model, to the point of dragging it doggedly into a future for which it is manifestly unfit?

But perhaps something is changing. In 2017, an American college sophomore named Trevor Hill stood up during a televised town hall meeting in New York and posed a simple question to Nancy Pelosi, the Speaker of the US House of Representatives at the time and one of the most powerful people in the world. He cited a study by Harvard University showing that 51% of US Americans between the ages of eighteen and twenty-nine no longer support capitalism, and asked whether the Democrats,

Pelosi's party, could embrace this fast-changing reality and stake out a vision for an alternative economy.[39]

Pelosi was visibly taken aback. 'I thank you for your question,' she said, 'but I'm sorry to say we're capitalists, and that's just the way it is.'

The footage went viral. It was powerful because it dramatised the taboo against questioning capitalism, right out in the open. Trevor Hill is no hardened left-winger. He's just your average millennial – bright, informed, curious about the world, and eager to imagine a better one. He had asked a sincere question, and yet Pelosi, stammering and defensive, was unable to entertain it, and unable even to articulate a meaningful justification for her position. Capitalism is so taken for granted that its proponents don't even know how to justify it. Pelosi's response – 'That's just the way it is' – was intended to shut down the question. But it did the opposite. It exposed the frailty of a tired ideology. It was like pulling back the curtain on the Wizard of Oz.

The video captured people's imaginations because it revealed that younger people are ready to think differently; ready to question old certainties. And they are not alone. While most people may not describe themselves as anti-capitalist, survey results nonetheless show that large majorities question core tenets of capitalist economics. A YouGov poll in 2015 found that 64% of people in Britain believe capitalism is unfair. Even in the US, it's as high as 55%. In Germany, a solid 77%. In 2020, a survey by the Edelman Trust Barometer showed that a majority of people around the world (56%) agree with the statement, 'Capitalism does more harm than good'. In France it's as high as 69%. In India it's a staggering 74%.[40] On top of this, fully three-quarters of people across all major capitalist economies say they believe corporations are corrupt.[41]

These sentiments become even stronger when the questions are framed in terms of growth. A poll conducted by Yale University in 2018 found that no fewer than 70% of Americans agree with the statement that 'environmental protection is more important than growth'. And these results hold even in Republican states, including in the deep South. The results are lowest in Oklahoma, Arkansas and West Virginia, but even there an overwhelming majority of voters (64%) take this position.[42] This completely upends longstanding assumptions about American attitudes towards the economy.

In 2019, the European Council on Foreign Relations asked an even stronger version of this question to people in fourteen EU countries. They phrased it as: 'Do you believe that environment should be made a priority even if doing so *damages* economic growth?' Surely people would be hesitant to agree with this kind of trade-off. Yet in almost all cases, large majorities (between 55% and 70%) said yes. There were only two exceptions, where support fell just shy of 50%. We find similar results outside Western Europe and North America. A scientific review of surveys found that when people have to choose between environmental protection and growth, 'environmental protection is prioritised *in most surveys and countries*'.[43]

In some surveys, it's clear that people are willing to go further still. A major consumer research study found that on average about 70% of people in middle- and high-income countries around the world believe that over-consumption is putting our planet and society at risk, that we should buy and own *less*, and that doing so would not compromise our happiness or well-being.[44] These are striking results. However these people might describe their political views, they are articulating principles that run directly counter to the core logic of capitalism. This is an extraordinary story that has been almost completely hidden from view. People around the world are yearning, quietly, for something better.

Degrowth

Sometimes scientific evidence conflicts with the dominant world view of a civilisation. When that happens, we have to make a choice. Either we ignore science, or we change our world view. When Charles Darwin first proved that all species, including humans, were descended from common ancestors over deep time, he was laughed off the stage. The notion that humans evolved from non-humans, instead of being created in the image of God; and the notion that the history of life on this planet stretches back much further than the few thousand years the Bible seems to suggest – at the time these ideas were utterly unacceptable. Some tried to explain Darwin's evidence away by devising outlandish alternative theories, in a desperate attempt to preserve the status quo. But the cat was out of the bag. Before long, Darwin's work had become scientific consensus, and it forever changed the way we see the world.

Something similar is happening right now. As evidence about the relationship between GDP growth and ecological breakdown continues to mount, scientists around the world are shifting their approach. In 2018, 238 scientists called on the European Commission to abandon GDP growth and focus on human well-being and ecological stability instead.[45] The following year, more than 11,000 scientists from over 150 countries published an article calling on the world's governments 'to shift from pursuing GDP growth and affluence toward sustaining ecosystems and improving well-being.'[46] This would have been unthinkable in mainstream circles only a few years ago, but now there's a striking new consensus forming.

Moving away from growth is not as wild as it might seem. For decades we've been told that we need growth in order to improve people's lives. But it turns out this isn't actually true. Beyond a certain point, which high-income countries have long surpassed, the relationship between GDP and social outcomes begins to break down. This should not be particularly surprising. GDP is an indicator of aggregate production, as measured in terms of real market prices. As we'll see in Chapter 4, it's not *increasing aggregate production* that matters; what matters is *what* we are producing, whether people have access to the things they need to live decent lives, and how income is distributed. The question of distribution is particularly important here, because right now income is distributed very, very unequally. Consider this: the richest 1% (all of whom are millionaires) capture some $19 trillion in income every year, which represents nearly a quarter of global GDP.[47] This is astonishing, when you think about it. It means that a quarter of all the labour we render, all the resources we extract, and all the CO_2 we emit is done to make rich people richer.

High-income countries don't need more growth in order to improve people's lives. What they need is to organise the economy around human well-being, rather than around capital accumulation. Once we realise this, it frees us to think much more rationally about how to respond to the crisis we face. Scientists have made it clear that the only feasible way to reverse ecological breakdown and keep global warming under 1.5°C, or even 2°C, is for high-income countries to actively reduce excess resource use and energy use.[48] Reducing resource use removes pressure from ecosystems and gives the web of life a chance to knit itself back together, while reducing energy use makes it much easier for us to accomplish a rapid transition to renewables – in a matter of years, not decades – before dangerous tipping points begin to cascade. How can we accomplish this? In a post-growth economy, some of it can be won through efficiency improvements. But we also need to scale down less necessary forms of production.

This is called 'degrowth' – a planned reduction of excess energy and resource use to bring the economy back into balance with the living world in a safe, just and equitable way.[49] The exciting part is that we know we can do this while at the same time ending poverty, improving human well-being, and ensuring good lives for all.[50] Indeed, that is the core principle of degrowth.

What does this look like in practice? It's really quite straightforward. Right now, the dominant assumption in economics is that all sectors of the economy must grow, all the time, regardless of whether or not we actually need them to. This is an irrational way to manage an economy at the best of times, but during an ecological emergency it is clearly dangerous. Instead, we should decide what kinds of things we *need* to increase (things like clean energy, public healthcare, essential services, regenerative agriculture – you name it), and what sectors are less necessary, or ecologically destructive, and should be radically reduced (things like fossil fuels, private jets, arms and SUVs). We can also scale down forms of production that are designed purely to maximise profits rather than to meet human needs, like planned obsolescence, where products are made to break down after a short time, or advertising strategies intended to manipulate our emotions and make us feel that what we have is inadequate.

As we slow down excess production and liberate people from the toil of unnecessary labour, we can shorten the working week to maintain full employment, distribute income and wealth more fairly, and expand access to key public services like universal healthcare, education and affordable housing. As we'll see in Chapter 5, these measures have been proven, over and over again, to have a powerful positive impact on people's health and well-being. These are the keys to a flourishing society, and they allow us to delink social progress from economic growth. The evidence is truly inspiring.

Let me emphasise that degrowth is *not* about reducing GDP. GDP is not a dial we can turn. Of course, slowing down unnecessary production, and decommodifying public services, is likely to cause GDP to grow more slowly, or stop growing, or even decline. And if so, that's OK. Under normal circumstances, this might trigger a recession. But a recession is what happens when a growth-dependent economy stops growing. It is chaotic and disastrous. What I'm calling for here is something completely different. It is about shifting to a different kind of economy altogether – an economy that doesn't *require* growth in the first place. To get there, we need to rethink everything from the debt system to the banking system, to liberate people, businesses, states and even innovation itself from the stuffy constraints of the growth imperative, freeing us to focus on higher goals.

As we take practical steps in this direction, exciting new possibilities come into view. We can create an economy that is organised around human flourishing instead of around endless capital accumulation; in other words, a *post-capitalist economy.* An economy that's fairer, more just, and more caring.

These ideas have been percolating on different continents for the past few decades, like whispers of hope. We inherit them from people like Herman Daly and Donella Meadows, the pioneering founders of ecological economics; from philosophers like Vandana Shiva and André Gorz; social scientists like Arturo Escobar and Maria Mies; economists like Serge Latouche and Giorgos Kallis; and from Indigenous writers and activists like Ailton Krenak and Berta Cáceres.[51] Suddenly these ideas are rushing into the mainstream, and inspiring an extraordinary shift in scientific discourse. Now we have a choice before us: will we ignore science in order to maintain our world view, or will we change our world view? This time the stakes are much higher than they were in Darwin's age. This time we don't have the luxury of pretending the science doesn't exist. This time, it's a matter of life and death.

*

To find the path ahead of us, we first need to understand how we got locked into the growth imperative to begin with. This requires reaching into the deep history of capitalism, to grasp the inner logic of how it works, and how it came to be imposed around the world – a journey we begin in Chapter 1. Along the way, we will discover that there is something else at stake; something unexpected. The processes of extraction that are so central to capitalist growth ultimately depend on a particular kind of ontology, or theory of being. Indeed, this is where our problem ultimately lies.

Those of us who live in capitalist societies today have been taught to believe that there is a fundamental distinction between human society and the rest of the living world: humans are separate from and superior to 'nature'; humans are subjects with spirit and mind and agency, whereas nature is an inert, mechanistic object. This way of seeing the world is known as dualism. We inherit these ideas from a long line of thinkers, from Plato to Descartes, who primed us to believe that humans can rightfully exploit nature and subject it to our control. We didn't always believe these things. In fact, those who sought to pave the way for capitalism in the sixteenth century first had to destroy other, more holistic ways of seeing the world, and either convince or force people to become dualists. Dualist philosophy was leveraged to cheapen life for the sake of growth; and it is responsible at a deep level for our ecological crisis.

But this is not the only way of being that's available to us. My colleagues in anthropology have long pointed out that for most of human history people operated with a very different ontology – a theory of being that we refer to, broadly, as animist. For the most part, people saw no fundamental divide between humans and the rest of the living world. Quite the opposite: they recognised a deep interdependence with rivers, forests, animals and plants, even with the planet itself, which they saw as sentient beings, just like people, and animated by the very same spirit. In some cases they even regarded them as kin.

We see traces of this philosophy still flourishing today, from the Amazon basin to the highlands of Bolivia to the forests of Malaysia, where people think about and interact with non-human beings – from jaguars to rivers – not as 'nature' but as relatives. When you see the world this way, it fundamentally changes how you behave. If you start from the premise that all beings are the moral equivalent of persons, then you cannot simply take from them. To exploit nature as a 'resource' for the sake of human enrichment is morally reprehensible – similar to slavery or even to cannibalism. Instead, you have to enter into a relationship of reciprocity, in the spirit of the gift. You have to give at least as much as you receive.

This logic, which has inherent ecological value, runs directly against the core logic of capitalism, which is to take – and, more importantly, to take more than you give back. In fact, as we will see, this is the basic mechanism of growth.

Enlightenment thinkers once disparaged animist ideas as backwards and unscientific. They considered them to be a barrier to capitalist expansion, and sought desperately to stamp them out.

But today science is beginning to catch up. Biologists are discovering that humans are not standalone individuals, but composed largely of microorganisms on which we depend for functions as basic as digestion. Psychiatrists are learning that spending time around plants is essential to people's mental health, and indeed that certain plants can heal humans from complex psychological traumas. Ecologists are learning that trees, far from being inanimate, communicate with each other and even share food and medicine through invisible mycelial networks in the soil. Quantum physicists are teaching us that individual particles that appear to be distinct are inextricably entangled with others, even across vast distances. And Earth-systems scientists are finding evidence that the planet itself operates like a living superorganism.

All of this is changing how we think about our position in the web of life, and paving the way for new theories of being. At the very time our planet is plunging into ecological catastrophe, we are beginning to learn a different way of seeing ourselves in relation to the rest of the living world. We are beginning to remember secrets we long ago forgot; secrets that linger in our hearts like whispers from the ancestors.

This completely upends the dusty old tropes of twentieth-century environmentalism. Environmentalists sometimes have a tendency to speak in terms of 'limits', meagreness and personal puritanism. But this gets it exactly back to front. The notion of limits puts us on the wrong foot from the start. It presupposes that nature is something 'out there', separate from us, like a stern authority hemming us in. This kind of thinking emerges from the very dualist ontology that got us into trouble in the first place. What I am calling for here is something altogether different. It is not about limits but interconnectedness – recovering a radical intimacy with other beings. It is not about

33

puritanism but pleasure, conviviality and fun. And it is not about meagreness but bigness – expanding the boundaries of human community, expanding the boundaries of our language, expanding the boundaries of our consciousness.[52]

It's not just our economics that needs to change. We need to change the way we see the world, and our place within it.

Glimpses of a future

Sometimes new ideas can make you see everything differently. Old myths fall apart, and new possibilities come into view. Difficult problems melt away, or become much easier to solve. Things that once seemed unthinkable suddenly become obvious. Whole worlds can change.

I like to imagine a time in the future when I'm again captivated by the number of insects back home in Eswatini. I'm an old man, sitting on the porch in the evening, watching them in awe, listening to their chirping, just as I did as a child. In this vision, a lot has changed about the world. High-income countries brought their use of resources and energy down to sustainable levels. We began to take democracy seriously, shared income and wealth more fairly, and put an end to poverty. The gap between rich countries and poor countries shrank. The word 'billionaire' disappeared from our languages. Working hours fell from forty or fifty hours a week down to twenty or thirty, giving people more time to focus on community, caring and the arts of living. High-quality public healthcare and education were made available to everyone. People came to live longer, happier, more meaningful lives. And we began to think of ourselves differently: as beings interconnected with, rather than separate from, the rest of the living world.

As for the planet, something remarkable happened. The rainforests grew back, across the Amazon, the Congo and Indonesia; dense and green and teeming with life. Temperate forests spread again across Europe and Canada. Rivers ran clear, and filled with fish. Whole ecosystems recovered. We accomplished a quick transition to renewable energy, global temperatures stabilised, and weather systems began to return to their ancient patterns. In a word, things started to heal . . . *we* began to heal . . . and faster than anybody imagined was possible. We took less, but we gained so much more.

This book is about that dream. We have a journey ahead, which will carry us over 500 years of history. We'll explore the roots of our current economic system, how it took hold, and what makes it tick. We will look at concrete, practical steps we can take to reverse ecological breakdown and build an alternative, post-capitalist economy. And we will travel across continents, to cultures and communities that interact with the living world in ways that open up whole new horizons of the imagination.

Right now it may just be the faintest whisper of a possibility. But whispers can build into winds, and take the world by storm.

PART ONE

More is Less

One

Capitalism: A Creation Story

> Animism had endowed things with souls; industrialism makes souls into things.
>
> Max Horkheimer and Theodor Adorno

We humans have been on this planet for nearly 300,000 years; fully evolved, fully intelligent, exactly as we are today. For approximately 97% of that time our ancestors lived in relative harmony with the Earth's ecosystems. This is not to say that early human societies didn't change ecosystems, and it's not to say there weren't problems. We know, for example, that certain societies played a role in the demise of some of the planet's ancient megafauna, like woolly mammoths and giant sloths and sabre-toothed cats. But they never precipitated anything like the multi-front ecological catastrophe that we are witnessing today.

It was only with the rise of capitalism over the past few hundred years, and the breathtaking acceleration of industrialisation from the 1950s, that on a planetary scale things began to tip out of balance. Once we understand this, it changes how we think about the problem. We call this human epoch the Anthropocene, but in fact this crisis has nothing to do with humans *as such*. It has to do with the dominance of a particular economic system: one that is recent in origin, which developed in particular places at a particular time in history, and which has not been adopted to the same extent by all societies. As the sociologist Jason Moore has pointed out, this isn't the Anthropocene – it's the Capitalocene.[1]

This can be difficult to wrap our minds around at first. We tend to take capitalism so much for granted that we just assume it has more or less always been around, at least in nascent form; after all, capitalism is about markets, and markets are ancient. But this is a false equivalence. While markets have been around for many thousands of years, in different times and places, capitalism is relatively recent – only about 500 years old.[2] What makes capitalism distinctive isn't that it has markets, but that it is organised around perpetual growth; indeed, it is the first intrinsically expansionist economic system in history. It pulls ever-rising quantities of nature and human labour into circuits of commodity production. And because the goal of capital is to extract and accumulate surplus, it has to get these things for as cheap as possible. In other words, capital works according to a simple, straightforward formula: take more – from nature and from labour – than you give back.

The ecological crisis is an inevitable consequence of this system. Capitalism has tipped us out of balance with the living world. Once we grasp this fact, new questions come rushing to mind: How did this happen? Where did capitalism come from? Why did it take hold?

The usual story holds that it's in our 'nature' to be self-interested, maximising agents – what some have described as *homo economicus* – the profit-seeking automatons that we encounter in microeconomics textbooks. We are taught that this natural tendency gradually broke through the constraints of feudalism, put an end to serfdom, and gave rise to capitalism as we know it today. That's our story. It is our Origin Tale. It gets repeated so often that everyone just accepts it. And because the rise of capitalism is cast as an expression of innate human nature – human selfishness and greed – problems like inequality and ecological breakdown seem inevitable and virtually impossible to change. But, remarkably for a story that has become so entrenched in our culture, none of this is true. Capitalism didn't just 'emerge'. There was no smooth, natural 'transition' to capitalism, and it has nothing to do with human nature. Historians have a much more interesting and significantly darker story to tell – a story that reveals some surprising truths about how our economy actually works. Understanding this story helps us grasp the deep drivers of the ecological crisis, and offers important clues as to what we can do about it.

A forgotten revolution

Everyone learns in school that feudalism was a brutal system that produced terrible human misery. And it's true. Lords and nobles controlled the land, and the people who lived on it – serfs – were forced to render tribute to them in the form of rents, taxes, tithes and unpaid labour. But contrary to our dominant narratives, it wasn't the rise of capitalism that put an end to this system. That victory belongs, remarkably enough, to a courageous struggle fought by a long tradition of everyday revolutionaries who have for some reason been almost entirely forgotten.

In the early 1300s, commoners across Europe began rebelling against the feudal system. They refused to submit to unpaid labour, they rejected the taxes and tithes extracted by lords and the Church, and they began demanding direct control over the land they tilled. These were not just petty complaints popping up here and there. It was organised resistance. And in some cases it grew into outright military conflicts. In 1323, peasants and workers took up arms in Flanders in a battle that lasted five years before their defeat by the Flemish nobility. Similar rebellions erupted elsewhere across Europe – in Bruges, Ghent, Florence, Liège and Paris.[3]

These early rebellions had little success. In most cases they were crushed by well-armed militaries. And when the Black Death struck in 1347 things only seemed to get worse: bubonic plague wiped out a third of Europe's population, triggering an unprecedented social and political crisis.

But in the wake of this disaster, something unexpected happened. Because labour was scarce and land abundant, suddenly peasants and workers had more bargaining power. They were able to demand lower rents for land, and higher wages for their labour. Lords found themselves on the back foot, and the balance of power tilted in commoners' favour for the first time in generations. Commoners began to realise that this was their chance: they had an opportunity to change the very foundations of the social and political order. They grew more hopeful, more confident, and the rebellions gained steam.[4]

In England, Wat Tyler led a peasants' revolt against feudalism in 1381, inspired by the radical preacher John Ball, famous for his call: 'Now the time is come in which ye may (if ye will) cast off the yoke of bondage, and recover liberty.' In 1382 a revolt in the Italian city of Ciompi succeeded in taking over the government. In Paris, a 'workers' democracy' seized power in 1413. And in 1450 an army of English peasants and workers marched on London in what became known as Jack Cade's Rebellion. Entire regions rose up during this period, forming assemblies and recruiting armies.

By the middle of the 1400s, wars were erupting between peasants and lords across Western Europe, and as the rebels' movement grew their demands broadened. They weren't interested in tweaking the system a bit around the edges – they wanted nothing short of revolution. According to the historian Silvia Federici, an expert in the political economy of the Middle Ages, 'the rebels did not content themselves with demanding some restrictions to feudal rule, nor did they only bargain for better living conditions. Their aim was to put an end to the power of the lords.'[5]

While in most cases the individual rebellions themselves were put down (Wat Tyler and John Ball were executed along with 1,500 of their followers), the movement ultimately succeeded in destroying serfdom across much of the continent. In England, the practice was almost completely eradicated in the wake of the 1381 revolt. Serfs became free farmers, subsisting on their own lands, with free access to commons: pasture for grazing, forests for game and timber, waterways for fishing and irrigation. They worked for wages if they wanted extra income – rarely under coercion. In Germany, peasants came to control up to 90% of the country's land. And even where feudalistic relations remained intact, conditions for peasants improved significantly.

As feudalism fell apart, free peasants began to build a clear alternative: an egalitarian, co-operative society rooted in the principles of local self-sufficiency. The results of this revolution were astonishing, in terms of the welfare of commoners. Wages rose to levels higher than ever before in recorded history, doubling or even tripling in most regions and in some cases rising as much as sixfold.[6] Rents declined, food became cheap, and nutrition improved. Workers were able to bargain for shorter working hours and weekends off, plus benefits like meals on the job and payment for each mile they had to travel to and from work. Women's wages shot up too, narrowing what under feudalism had been a substantial gender pay gap. Historians have described the period from 1350 to 1500 as 'the golden age of the European proletariat'.[7]

It was a golden age for Europe's ecology, too. The feudal system had been an ecological disaster. Lords put peasants under heavy pressure to extract from the land and forests while giving nothing back. This drove a crisis of deforestation, overgrazing, and a gradual decline in soil fertility. But the political movement that emerged after 1350 reversed these trends and inaugurated a period of ecological regeneration. Once they won direct control of the land, free peasants were able to maintain a more reciprocal relationship with nature: they managed pastures and commons collectively, through democratic assemblies, with careful rules that regulated tillage, grazing and forest use.[8] Europe's soils began to recover. The forests regrew.

Backlash

Needless to say, Europe's elites were not pleased with this turn of events. They considered the high wages 'scandalous', and were irritated that commoners would only hire themselves out for short periods or limited tasks, leaving as soon as they had enough income to satisfy their needs. 'Servants are now masters and masters servants,' complained John Gower in *Miroir de l'Omme* (1380). As one writer put it in the early 1500s: 'The peasants are too rich . . . and do not know what obedience means; they don't take law into any account, they wish there were no nobles . . . and they would like to decide what rent we should get for our lands.'[9] And according to another: 'The peasant pretends to imitate the ways of the freeman, and gives himself the appearance of him in his clothes.'[10]

During the revolutionary period from 1350 to 1500, elites suffered what historians have described as a crisis of 'chronic disaccumulation'.[11] As national income was shared more evenly across the population it became more difficult for elites to pile up the profits they had enjoyed under feudalism. This is an important point. We often assume that capitalism emerged somehow naturally from the collapse of feudalism, but in fact such a transition would have been impossible. Capitalism requires elite accumulation: piling up excess wealth for large-scale investment. But the egalitarian conditions of post-feudalist society – self-sufficiency, high wages, grassroots democracy and collective management of resources – were inimical to the possibility of elite accumulation. Indeed, this is exactly what elites were complaining about.

What that new society might have grown to look like we will never know, for it was brutally crushed. Nobles, the Church and the merchant bourgeoisie united in an organised attempt to end peasant autonomy and drive wages back down. They did so not by re-enserfing peasants – that had proved to be impossible. Rather, they forced them off their land in a violent, continent-wide campaign of evictions. As for the commons – those collectively managed pastures, forests and rivers that sustained rural communities – they were fenced off and privatised for elite use. They became, in a word, *property*.

This process was known as 'enclosure'.[12] Thousands of rural communities were destroyed during the enclosure movement; crops were ripped up and burned, whole villages razed to the ground. Commoners lost their access to land, forests, game, fodder, water, fish – all the resources necessary for life. And the Reformation added further fuel to the bonfire of dispossession: as Catholic monasteries were dismantled across Europe, their lands were snapped up by nobles and cleared of the people who lived there.

Peasant communities didn't go down without a fight, of course. But they had precious little success. In Germany, an organised peasant rebellion in 1525 was defeated in a massacre that left more than 100,000 commoners dead – one of the bloodiest slaughters in world history. In 1549, a rebellion led by Englishman Robert Kett managed to take control of Norwich, the country's second largest city, before the military put them down: 3,500 rebels were massacred and their leaders hanged from the city's walls. A rebellion known as the Midland Revolt in 1607 culminated in an insurrection at Newton, where peasants ended up yet again in armed combat with enclosers. Fifty were executed in the defeat that followed.

Over the course of three centuries, huge swathes of Britain and the rest of Europe were enclosed and millions of people removed from the land, triggering an internal refugee crisis. It would be difficult to overstate the upheaval that characterised this period – it was a humanitarian catastrophe. For the first time in history, commoners were systematically denied access to the most basic resources necessary for survival. People were left without homes and food. We don't need to romanticise subsistence life to recognise that enclosure produced conditions that were far worse; worse even than under serfdom. In England, the word 'poverty' came into common use for the first time to describe the mass of 'paupers' and 'vagabonds' that enclosure produced – words that prior to this period rarely if ever appeared in English texts.

Yet as far as Europe's capitalists were concerned, enclosure was working like magic. It enabled them to appropriate huge amounts of land and resources that had previously been off limits. Economists have always recognised that some kind of initial accumulation was necessary for the rise of capitalism. Adam Smith called this 'previous accumulation', and claimed that it came about because a few people worked really hard and saved their earnings – an idyllic tale that still gets repeated in economics textbooks. But historians see it as naïve. This was no innocent process of saving. It was a process of plunder. Karl Marx insisted on calling it 'primitive accumulation', to highlight the barbaric nature of the violence it entailed.

But the rise of capitalism also depended on something else. It needed labour. Lots of it, and cheap. Enclosure solved this problem too. With subsistence economies destroyed and commons fenced off, people had no choice but to sell their labour for wages – not to earn a bit of extra income, as under the previous regime, nor to satisfy the demands of a lord, as under serfdom, but *simply in order to survive*. They became, in a word, proletarians. This was utterly new in world history. Such people were referred to at the time as 'free labourers', but this term is mis-leading: true, they were not forced to work as slaves or serfs, but they nonetheless had little choice in the matter, as their only alternative was starvation. Those who controlled the means of production could get away with paying rock-bottom wages, and people would have to take it. Any wage, no matter how small, was better than death.

*

All of this upends the usual story that we're told about the rise of capitalism. This was hardly a natural and inevitable process. There was no gradual 'transition', as people like to assume, and it certainly wasn't peaceful. Capitalism rose on the back of organised violence, mass impoverishment, and the systematic destruction of self-sufficient subsistence economies. It did not put an end to serfdom; rather, it put an end to the progressive revolution that had ended serfdom. Indeed, by securing virtu-ally total control over the means of production, and rendering peasants and workers dependent on them for survival, capital-ists took the principles of serfdom to new extremes. People did not welcome this new system with open arms; on the contrary, they rebelled against it. The period from 1500 to the 1800s, right into the Industrial Revolution, was among the bloodiest, most tumultuous times in world history.

For human welfare, the consequences of enclosure were devas-tating. It reversed all of the gains the free peasants had won. According to the economists Henry Phelps Brown and Sheila Hopkins, from the 1500s to the 1700s real wages declined by as much as 70%.[13] Nutrition deteriorated and starvation became commonplace: some of the worst famines in European history struck in the 1500s, as subsistence economies were ripped up.

The social fabric was left so shredded that between 1600 and 1650 populations across Western Europe actually declined. In England, we can see the imprint of this catastrophe clearly in the historical public health record: average life expectancy at birth fell from forty-three years in the 1500s to the low thirties in the 1700s.[14]

We all know that famous quote by Thomas Hobbes, where he says that life in the 'state of nature' was 'nasty, brutish and short'. He wrote those words in 1651. We read Hobbes as describing a putative condition of misery that existed *before* capitalism; a problem that capitalism was supposed to solve. But exactly the opposite is true. The misery he described was created by the rise of capitalism itself. Indeed, at that time Europe was one of the poorest, sickest places in the world; at least for commoners.[15] And what Hobbes didn't know is that it was about to get worse.

The enclosure movement went further in Britain than anywhere else in Europe. The monarchy had initially sought to limit enclosure, worried about the social crises it was creating. But those limits were abolished after the Civil War of the 1640s and the so-called Glorious Revolution of 1688, when the bourgeoisie assumed control of Parliament and obtained the power to do more or less whatever they pleased. Wielding the full force of the state, they introduced a series of laws – the Parliamentary Enclosures – that set off a wave of dispossession faster and more far-reaching than anything that had come before. Between 1760 and 1870, some 7 million acres were enclosed by legal writ, about one-sixth of England. By the end of this period there was almost no common land left in the country.

This final, dark episode in the destruction of the English peasant system coincided exactly with the Industrial Revolution. The dispossessed poured desperate and shell-shocked into the cities, where they provided the cheap labour that fuelled the dark Satanic mills immortalised in the poetry of William Blake.

Industrial capitalism took off, but at extraordinary human cost. Simon Szreter, one of the world's leading experts on historical public health data, has shown that this first century of the Industrial Revolution was characterised by a striking deterioration in life expectancy, down to levels not seen since the Black Death in the fourteenth century. In Manchester and Liverpool, the two giants of industrialisation, life expectancy collapsed compared to non-industrialised parts of the country. In Manchester it fell to a mere twenty-five years. And it was not just in England; this same effect can be seen in every other European country where it has been studied. The first few hundred years of capitalism generated misery to a degree unknown in the pre-capitalist era.[16]

Growth as colonisation

Historians have made big strides in understanding how the rise of capitalism depended on enclosure. But too often this story ignores the patterns of primitive accumulation that were playing out at the same time beyond Europe's shores, as part of the very same process. Across the global South, nature and human bodies were enclosed to an extent that dwarfed what happened within Europe itself.

When Europeans began to colonise the Americas in the decades after 1492, they were not driven by the romance of 'exploration' and 'discovery', as our schoolbooks would have it. Colonisation was a response to the crisis of elite disaccumulation that had been caused by the peasant revolutions in Europe. It was a 'fix'. Just as elites turned to enclosure at home, they sought new frontiers for appropriation abroad, beginning with Christopher Columbus's first voyage to the Americas. These two processes unfolded simultaneously. In 1525, the very year that German nobles massacred those 100,000 peasants, the Spanish king Carlos I awarded the kingdom's highest honour to Hernán Cortés, the conquistador who slayed 100,000 Indigenous people as his army marched through Mexico and destroyed the Aztec capital of Tenochtitlán. The congruence of these two events is no accident. In the decades that inaugurated the rise of capitalism, enclosure and colonisation were deployed as part of the same strategy.

The scale of colonial appropriation was staggering. From the early 1500s through the early 1800s, colonisers siphoned 100 million kilograms of silver out of the Andes and into European ports. To get a sense of the scale of this wealth, consider this thought experiment: if invested in 1800 at the historical average rate of interest, that quantity of silver would today be worth $165 trillion – more than double the world's GDP. And that's on top of the gold that was extracted from South America during the same period. This windfall played a key role in the rise of European capitalism. It provided some of the surplus that ended up invested in the Industrial Revolution; it enabled the purchase of land-based goods from the East, which allowed Europe to shift its population from agriculture to industrial production; and it funded the military expansion that powered further rounds of colonial conquest.[17]

Colonisation also provided the key raw materials that fuelled the Industrial Revolution. Take cotton and sugar, for example. Cotton was the most important commodity in Britain's industrial rise; the lifeblood of Lancashire's iconic mills. And sugar became a key source of cheap calories for Britain's industrial workers. But neither cotton nor sugar grow in Europe. To get them, Europeans appropriated vast tracts of land for plantation agriculture – millions of acres across much of Brazil, the West Indies and North America. By the year 1830, Britain alone was appropriating the equivalent of 25 to 30 million acres of productive land from its New World colonies.[18] And this was not regulated, conscientious extraction. Colonial mining, logging, and plantation monoculture caused ecological damage on a scale that was, up to that point, historically unprecedented. Indeed, what made the colonial frontiers so attractive to capital in the first place was that the land – and the people who lived on it – could be mistreated with impunity.

As for who powered all these mines and plantations: up to 5 million Indigenous Americans were enslaved for this purpose – a process so violent that it wiped out much of the population.[19] But even this was not enough. Another 15 million souls were shipped across the Atlantic from Africa during three centuries of state-sponsored human trafficking by European powers, from the 1500s to the 1800s. The United States extracted so much labour from enslaved Africans that, if paid at the US minimum wage, with a modest rate of interest, it would add up to $97 trillion today – four times the size of the US GDP.[20] And that's just the United States; it doesn't count the Caribbean and Brazil. The slave trade amounted to an extraordinary appropriation of labour, transferred from Indigenous and African communities into the pockets of European industrialists.

But there were also subtler forms of appropriation at work. In India, British colonisers used an insidious taxation system to extract an extraordinary quantity of goods and resources from the country's farmers and artisans. Between the years 1765 and 1938, they siphoned sums that would today be worth around $45 trillion out of India and into British coffers. This flow allowed Britain to buy strategic materials like iron, tar and timber, which were essential to the country's industrialisation. They also used it to finance the industrialisation of white settler colonies like Canada and Australia, and to pay for the British welfare system that, after the 1870s, finally started to address the misery generated by enclosure (in the late 19th century, more than half of Britain's domestic budget was funded by appropriation from India and other colonies).[21] Today, British politicians often seek to defend colonialism by claiming that Britain helped 'develop' India. But in fact exactly the opposite is true: Britain exploited India to develop itself.

The point here is that the rise of capitalism in Europe – and Europe's Industrial Revolution – did not emerge *ex nihilo*. It hinged on commodities that were produced by enslaved workers, on lands stolen from colonised peoples, and processed in factories staffed by European peasants who had been forcibly dispossessed by enclosure. We tend to think of these as separate processes, but they were all part of the same project, and operated with the same underlying logic. Enclosure was a process of internal colonisation, and colonisation was a process of enclosure. Europe's peasants were dispossessed from their lands just as Indigenous Americans were (although, notably, the latter were treated much worse, excluded from the realm of rights, and even humanity, altogether). And the slave trade is nothing if not the enclosure and colonisation of bodies – bodies that were appropriated for the sake of surplus accumulation just as land was, and treated as property in the same way.

It might be tempting to downplay these moments of violence as mere aberrations in the history of capitalism. But they are not. They are the foundations of it. Under capitalism, growth always requires new frontiers from which to extract uncompensated value. It is, in other words, intrinsically colonial in character.

Colonial interventions added a final piece to the rise of capitalism. Europe's capitalists had created a system of mass production, but they needed somewhere to sell. Who would absorb all this output? The enclosures provided a partial solution: by destroying self-sufficient economies, they created not only a mass of workers but also a mass of consumers – people wholly dependent on capital for food, clothes and other essential goods. But this alone was not enough. They needed to break into new markets abroad. The problem was that much of the global South, particularly Asia, had their own artisanal industries – often regarded as the finest in the world – and they were uninterested in importing things they could make for themselves. Colonisers solved this problem by using asymmetric trade rules to destroy local industries across the South, forcing the colonies to serve not only as a source of raw materials but also as a captive market for Europe's mass-produced goods. This completed the circuit. But the consequences were devastating: as European capital grew, the South's share of global manufacturing collapsed, from 77% in 1750 down to 13% by 1900.[22]

The paradox of artificial scarcity

In the wake of enclosure, Europe's peasants – those who remained in rural areas rather than migrating to cities – found themselves subject to a new economic regime. They were back once again under the rule of landlords, but this time in an even worse position: at least under serfdom they had secure access to land; now they were granted only temporary leases on it. And these weren't just ordinary leases. They were allocated on the basis of productivity. So to retain their access to land peasants had to devise ways to intensify their production, working longer hours and extracting more from the soil each year. Those who fell behind in this race would lose their tenancy rights and face starvation. This put peasants in direct competition with one another, with their own kin and neighbours, transforming what had been a system of collective co-operation into one organised around desperate antagonism.

The application of this logic to land and farming marked a fundamental transformation in human history. It meant that, for the first time, people's lives were governed by the imperatives of intensifying productivity and maximising output.[23] No longer was production about satisfying needs, no longer about local sufficiency; instead, it was organised around profit, and for the benefit of capital. This is crucial: those principles of *homo economicus* that we assume to be engraved in human nature were *instituted* during the enclosure process.[24]

The same pressures were at play in the cities. Refugees from enclosure who ended up in urban slums had no choice but to accept work for meagre wages. Because the refugees were many and jobs were few, competition among workers drove down the cost of labour, destroying the guild system that had previously protected the livelihoods of skilled craftsmen. Faced with the constant threat of replacement, workers were under pressure to produce as much as was physically possible; they regularly worked for sixteen hours a day, significantly longer than they had worked prior to enclosure.

These regimes of forced competition generated a dramatic surge in productivity. Between 1500 and 1900, the quantity of grain extracted per acre of land shot up by a factor of four. And it was this feature – known at the time as 'improvement' – that came to serve as the core justification for enclosure. The English land-owner and philosopher John Locke admitted that enclosure was a process of theft from the commons, and from commoners, but he argued that this theft was morally justifiable because it enabled a shift to intensive commercial methods that increased agricultural output.[25] Any increase in total output, he said, was a contribution to the 'greater good' – the betterment of humanity. The same logic was used to justify colonisation, and invoked by Locke himself to defend his claims to American lands. Improvement became the alibi for appropriation.

Today, the very same alibi is routinely leveraged to justify new rounds of enclosure and colonisation – of lands, forests, fisheries, of the atmosphere itself; but instead of 'improvement' we call it 'development', or 'growth'. Virtually anything can be justified if it contributes to GDP growth. We take it as an article of faith that growth benefits humanity as a whole; that it is essential to human progress. But even in Locke's time the alibi was clearly a ruse. While the commercialisation of agriculture did increase total output, the only 'improvement' was to the profits of the landowners. While output soared, commoners were hit by two centuries of increased famine. So too in the factories. None of the gains from the surge in labour productivity went back to the workers themselves; indeed, wages *declined* during the enclosure period. Profits were pocketed instead by those who owned the means of production.

The essential point to grasp here is that the emergence of the extraordinary productive capacity that characterises capitalism depended on creating and maintaining conditions of *artificial scarcity*. Scarcity – and the threat of hunger – served as the engine of capitalist growth. The scarcity was artificial in the sense that there was no actual deficit of resources: all the same land and forests and waters remained, just as they always had, but people's access to them was suddenly restricted. Scarcity was created, then, in the very process of elite accumulation. And it was enforced by state violence, with peasants massacred wherever they found the courage to tear down the barriers that cut them off from the land.[26]

*

This was a conscious strategy on the part of Europe's capitalists. In Britain, the historical record is full of commentary by landowners and merchants who felt that peasants' access to commons during the revolutionary period had encouraged them to leisure and 'insolence'. They saw enclosure as a tool for enhancing the 'industry' of the masses.

'Our forests and great commons make the poor that are upon them too much like the Indians,' wrote the Quaker John Bellers in 1695; '[they are] a hindrance to industry, and are nurseries of idleness and insolence'. Lord John Bishton, author of a 1794 report on agriculture in Shropshire, agreed: 'The use of common lands operates on the mind as a sort of independence.' After enclosure, he wrote, 'the labourers will work every day in the year, their children will be put out to labour early,' and 'that subordination of the lower ranks of society which in the present time is so much wanted would be thereby considerably secured.'

In 1771 the agriculturalist Arthur Young noted that 'everyone but an idiot knows that the lower classes must be kept poor, or they will never be industrious'. The Reverend Joseph Townsend emphasised in 1786 that 'it is only hunger which can spur and goad them on to labour'. 'Legal constraint,' Townsend went on, 'is attended with too much trouble, violence, and noise ... whereas hunger is not only a peaceable, silent, unremitted pressure, but as the most natural motive to industry, it calls forth the most powerful exertions ... Hunger will tame the fiercest animals, it will teach decency and civility, obedience and subjugation to the most brutish, the most obstinate, and the most perverse.'

Patrick Colquhoun, a powerful Scottish merchant, saw poverty as an essential precondition for industrialisation:

> Poverty is that state and condition in society where the individual has no surplus labour in store, or, in other words, no property or means of subsistence but what is derived from the constant exercise of industry in the various occupations of life. Poverty is therefore a most necessary and indispensable ingredient in society, without which nations and communities could not exist in a state of civilisation. It is the lot of man. It is the source of wealth, since without poverty, there could be no labour; there could be no riches, no refinement, no comfort, and no benefit to those who may be possessed of wealth.

David Hume (1752) built on these sentiments to elaborate an explicit theory of 'scarcity': 'Tis always observed, in years of scarcity, if it be not extreme, that the poor labour more, and really live better.'[27] These passages reveal a remarkable paradox. The proponents of capitalism themselves believed it was necessary to *impoverish* people in order to generate growth.

This same strategy was deployed across much of the rest of the world during European colonisation. In India, colonisers tried to pressure people to shift from subsistence farming to cash crops for export to Britain: opium, indigo, cotton, wheat and rice. But Indians were unwilling to make this transition voluntarily. To break their resistance, British officials imposed taxes that plunged peasants into debt, leaving them with no choice but to comply. The British East India Company and later the Raj sought to speed this transition along by dismantling the communal support systems that people relied on: they destroyed granaries, privatised the irrigation systems, and enclosed the commons that people used for wood, fodder and game. The theory was that these traditional welfare systems made people 'lazy', accustomed to easy food and leisure; by removing them, you could discipline people with the threat of hunger, and get them to compete with one another to extract ever higher yields from the land.

From the perspective of agricultural productivity, it worked; but the destruction of subsistence agriculture and communal support systems left peasants vulnerable to market fluctuations and droughts. During the last quarter of the nineteenth century, the height of the British Empire, 30 million Indians perished needlessly of famine in what the historian Mike Davis has called the 'Late Victorian Holocausts'. Needlessly, because even at the peak of the famine there was a net surplus of food. In fact, Indian grain exports more than tripled during this period, from 3 million tons in 1875 to 10 million tons in 1900. This was artificial scarcity taken to new extremes – far worse than anything that was inflicted within Europe. [28]

In Africa, colonisers faced what they openly called 'the Labour Question': how to get Africans to work in mines and on plantations for low wages. Africans generally preferred their subsistence lifestyles, and showed little inclination to do back-breaking work in European industries. The promise of wages was in most cases not enough to induce them into what they considered to be needless labour. Europeans fumed at this resistance, and responded by either forcing people off their land (the Native Lands Act in South Africa shoved the black population onto a mere 13% of the country's territory), or forcing them to pay taxes in European currency. Either course of action left Africans with no option but to sell themselves for wages.

The same process of enclosure and forced proletarianisation played out over and over again during the period of European colonisation – not just under the British but under the Spanish, Portuguese, French and Dutch as well – with examples too numerous to recite here. In all of these cases scarcity was created, purposefully, for the sake of capitalist expansion.

*

How odd that the history of capitalism – a system that generated such extraordinary material productivity – is marked by the constant creation of scarcity, scarred by devastating famines and a centuries-long process of immiseration. This apparent contradiction was first noticed in 1804 by James Maitland, the 8th Earl of Lauderdale.[29] Maitland pointed out that there is an inverse relation between what he called 'private riches' and 'public wealth', or commons, such that an increase in the former can only ever come at the expense of the latter.

'Public wealth,' Maitland wrote, 'may be accurately defined *to consist of all that man desires, as useful or delightful to him.*' In other words, it has to do with goods that have an intrinsic use value even when they are abundant, including air, water and food. Private riches, on the other hand; consist '*of all that man desires as useful or delightful to him; which exists in a degree of scarcity.*' The scarcer something is, the more money you can extort from people who need it. For instance, if you enclose an abundant resource like water and establish a monopoly over it, you can charge people to access it and therefore increase your private riches. This would also increase what Maitland called the 'sum-total of individual riches' – what today we might call GDP. But this can be accomplished only by curtailing people's access to what was once abundant and free. Private riches go up, but public wealth goes down. This became known as the 'Lauderdale Paradox'.

Maitland recognised that this was happening during the process of colonisation. He noticed that colonisers were burning down orchards that produced fruits and nuts, so people who once lived off the natural abundance of the land would be compelled instead to work for wages and purchase food from Europeans. What was once abundant had to be made scarce. Perhaps the most iconic example of this was the salt tax the British Raj imposed on India. Salt was freely available all along India's coasts – all you had to do was bend down and scoop it up. Yet the British made people pay for the right to do this, as part of a scheme to produce revenue for the colonial government. Public wealth had to be sacrificed for the sake of private riches; commons sabotaged for growth.

The great separation

Enclosure and colonisation were necessary preconditions for the rise of European capitalism. They opened frontiers for the appropriation of cheap resources, destroyed subsistence economies, created a mass of cheap labour, and by generating artificial scarcity set the engines of competitive productivity in motion. Yet, as powerful as these forces were, they were not sufficient to break down the barriers to elite accumulation. Something else was needed – something far subtler but nonetheless equally violent. Early capitalists not only had to find ways to compel people to work for them, they also had to change people's beliefs. They had to change how people regarded the living world. Ultimately, capitalism required a new story about nature.

*

For most of our 300,000-year history, we humans have had an intimate relationship with the rest of the living world. We know that people in early human societies were likely to be able to describe the names, properties and personalities of hundreds if not thousands of plants, insects, animals, rivers, mountains and soils, in much the same way people today know the most recondite facts about actors, celebrities, politicians and product brands. Aware that their existence depended on the well-being of other living systems around them, they paid close attention to how those systems worked. They regarded humans as an inextricable part of the rest of the living community, which they saw in turn as sharing the essential traits of humanity. Indeed, the art our ancestors left hidden on stone surfaces around the world suggests that they believed in a sort of spiritual interchangeability between humans and non-human beings.

Anthropologists refer to this way of seeing the world as animism – the idea that all living beings are interconnected, and share in the same spirit or essence. Because animists draw no fundamental distinction between humans and nature, and indeed in many cases insist on the underlying relatedness – even *kinship* – of all beings, they have strong moral codes that prevent them from exploiting other living systems. We know from animist cultures today that while people of course fish, hunt, gather and farm, they do so in the spirit not of extraction but of *reciprocity*. Just as with gifts exchanged among people, transactions with non-human beings are hedged about with rituals of respect and politeness. Just as we take care not to exploit our own relatives, so animists are careful to take no more than ecosystems can regenerate, and give back by protecting and restoring the land.

In recent years anthropologists have come to see this as more than just a cultural difference. It is deeper than that. It is a fundamentally different way of conceptualising the human. It is a different kind of *ontology* – an ontology of inter-being.

This ontology came under attack with the rise of empires, which gradually came to see the world as split in two, with a spiritual realm of gods separate from and above the rest of creation. Humans were given a privileged place in this new order: made in the image of the gods themselves, and thus possessed of the right to rule over the rest of creation. This idea – the principle of 'dominion' – grew firmer during the Axial Age with the rise of transcendental philosophies and religions across the major Eurasian civilisations: Confucianism in China; Hinduism in India; Zoroastrianism in Persia; Judaism in the Levant and Sophism in Greece. We can see it spelled out in ancient Mesopotamian texts dating back 3,000 years. And perhaps nowhere is this clearer than in Genesis itself:

> And God said, Let us make man in our image, after our likeness: and let them have dominion over the fish of the sea, and over the fowl of the air, and over the cattle, and over all the earth, and over every thing that creeps upon the ground.

In the fifth century BC this new way of seeing the world received a boost from Plato, who built his whole philosophy on the idea of a transcendental realm separate from an earthly realm. The transcendental realm was the source of abstract Truth and Reality, the ideal essence of things, while the material world was but a poor imitation – a mere shadow. This idea came to inform the Christian notion of a spiritual heaven set in opposition to a worldly realm of mere matter – sinful, decaying and passing away. Indeed, the Church, and the Christian Roman Empire that expanded across Europe, vigorously sponsored the Platonic view, which came to be formalised in the doctrine of *contemptus mundi*: 'contempt for the world'.

But despite the rise of these new ideas, most people held firm to relational ontologies. Even among philosophers, counter-discourses remained strong. Aristotle, Plato's most famous student, publicly rejected transcendentalism, insisting that the essence of things lies within them, not in some ethereal elsewhere, and that all beings have souls and share versions of the same spirit. Building on Aristotle, many philosophers regarded the living world itself as an intelligent organism, or even as a deity. Cicero wrote in the second century BC that 'the world is a living and wise being': it reasons and feels, and all its parts are interdependent. For the Stoics, who were influential in Athens during the first century, God and matter were synonymous – and therefore matter itself pulsed with divinity. The Roman philosopher Seneca saw the earth as a living organism with springs and rivers flowing through her like blood through veins, with metals and minerals forming slowly in her womb, and morning dew like perspiration on her skin.[30]

These ideas remained prominent in so-called pagan cultures across Europe, which rejected the Christian distinction between sacred and profane. They regarded the living world – plants and animals, mountains and forests, rivers and rain – as enchanted, filled with spirits and divine energy. As Christendom expanded through Europe it sought to repress these ideas wherever it encountered them, as in the persecution of the Celtic Druids, but it never succeeded in stamping them out; they remained common currency among peasants. In fact, after 1200 animistic ideas enjoyed a striking revival, as new translations of Aristotle's texts became available in Europe and gave legitimacy to peasant beliefs.[31] And in the wake of the peasant rebellions, as feudalism collapsed after 1350 and commoners wrested control of the land from feudal lords, these ideas became openly accepted.

We can trace animistic ontologies all the way to the Renaissance, where even then the dominant view regarded the material world as animated, and saw the Earth as a living, nurturing mother. In the fifteenth century, Pico della Mirandola wrote:

> All this great body of the world is a soul, full of intellect and of God, who fills it within and without and vivifies the All . . . The world is alive, all matter is full of life . . . Matter and bodies or substances . . . are energies of God. In the All there is nothing which is not God.

*

But then something happened. In the 1500s, there were two powerful factions of European society who were worried about the striking revival of animistic ideas, and set out to destroy them.

One was the Church. As far as the clergy were concerned, the notion that spirit suffused the material world threatened their claim to be the only conduits to the divine, and the only legitimate proxies of divine power. This was a problem not only for priests, but also for the kings and aristocrats who ultimately depended on their sanction. Animistic ideas had to be defeated because they were loaded with subversive implications. If spirit is everywhere, then there is no God – and if there is no God then there is no priest, and no king. In such a world, the divine right of kings crumbles into incoherence.[32] And that's exactly what happened. The ideas of Aristotle inspired many of the medieval peasant rebellions that sought to overthrow feudalism. These movements were denounced by the Church as heretical, and the charge of heresy was used to justify brutal violence against them.

But there was another powerful faction that regarded animist ideas as a problem: capitalists. The new economic system that began to dominate after 1500 required a new relationship with the land, with the soils, and with the minerals beneath the surface of the earth: one built on the principles of possession, extraction, commodification and ever-increasing productivity, or, in the discourse of the time, 'improvement'. But in order to possess and exploit something you must first regard it as an object. In a world where everything was alive and pulsing with spirit and agency, where all beings were regarded as subjects in their own right, this sort of possessive exploitation – in other words, property – was ethically unfathomable.

The historian Carolyn Merchant argues that animistic ideas limited the extent to which people considered it permissible to plunder the earth. 'The image of the earth as a living organism and nurturing mother had served as a cultural constraint restricting the actions of human beings,' she writes. 'One does not readily slay a mother, dig into her entrails for gold or mutilate her body . . . As long as the earth was considered to be alive and sensitive, it could be considered a breach of human ethical behaviour to carry out destructive acts against it.'[33]

This is not to say that people didn't extract from the land or mine the mountains. They did; but they did so with careful decorum and rituals of respect. Miners, smiths and farmers offered propitiation. They believed they were permitted to take from the earth, as one might receive a gift, but that to take too much, or too violently, would invite calamity. The Roman naturalist Pliny wrote in the first century that earthquakes were an expression of the earth's indignation at being mined out of avarice rather than out of need:

> We trace out all the veins of the earth, and yet . . . are astonished that it should occasionally cleave asunder or tremble: as though these signs could be any other than expressions of the indignation felt by our sacred parent! We penetrate her entrails, and seek for treasures . . . as though each spot we tread upon were not sufficiently bounteous and fertile for us!

Those who sought to advance capitalism had to find a way not only to strip humans from the land, but to destroy the animist ideas that enjoyed such prominence – to strip the earth of its spirit and render it instead a mere stock of 'natural resources' for humans to exploit.

*

They found their first answer in Francis Bacon (1561–1626), the Englishman celebrated as the 'father of modern science'.

Bacon's legacy is eulogised in school textbooks today, and for good reason: he made significant contributions to the scientific method. But there is a rather sinister side to his story that has largely fallen out of public consciousness. Bacon actively sought to destroy the idea of a living world, and to replace it with a new ethic that not only sanctioned but celebrated the exploitation of nature. To this end, he took the ancient theory of nature-as-female and transformed her from a nurturing mother into what he called a 'common harlot'. He cast nature, and indeed matter itself, as devious, disordered, wild and chaotic – a beast that, to quote his words, must be 'restrained', 'bound' and 'kept in order'.

For Bacon, science and technology were to serve as the instruments of domination. 'Science should as it were torture nature's secrets out of her,' Bacon wrote. And with the knowledge thus gained, 'man' would not 'merely exert a gentle guidance over nature's course', but 'have the power to conquer and subdue, to shake her to her foundations'. Nature must be 'bound into service' and made into a 'slave,' 'forced out of her natural state and squeezed and moulded' for human ends.

Bacon's use of torture as a metaphor here is revealing, as he himself – in his role as Attorney General under King James I – deployed torture against the peasant rebels and heretics of his time, and worked to legitimise the practice as a means of defending the state. Just as Bacon saw torture as a weapon against peasant insurrection, so he saw science as a weapon against nature. Like peasants, nature had resisted domination too long. Science was to break her once and for all.

In Bacon's writing we can also see hints of another idea emerging. Not only is nature something to be controlled and manipulated, it is also transformed from a living organism into inert matter. Nature may appear to be alive and moving, but its motion should be understood as that of a machine, Bacon said – nothing more than a system of pumps and springs and cogs. But it was in the hands of another man, only a few years later, that this vision of nature-as-machine was formulated into a coherent philosophy: the French thinker René Descartes.

Descartes realised that the domination of nature Bacon called for could only be justified if nature was rendered lifeless. To accomplish this, he reached back to Plato's idea of a world divided in two, and gave it a new spin. He argued that there was a fundamental dichotomy between mind and matter. Humans are unique among all creatures in having minds (or souls), he claimed – the mark of their special connection to God. By contrast, the rest of creation is nothing but unthinking material. Plants and animals have no spirit or agency, intention or motivation; they are mere automatons, operating according to predictable mechanical laws, ticking away like a clock (Descartes was famously enamoured of clocks).

In an attempt to prove the point, Descartes took to dissecting living animals. He nailed their limbs to boards and probed their organs and nerves – including, in one particularly grotesque episode, his wife's dog. While the animals writhed and wailed in agony, he insisted this was only the 'appearance' of pain, just a reflex: muscles and tendons responding automatically to physical stimuli. He urged people not to be fooled by the appearance of sentience or intelligence. It's not the deer or the owl itself that is the appropriate object of analysis, he said: to recognise the mechanical nature of life you have to dig in and peer at the parts, not the whole. What seems like life is really just inert matter. An object.

In Descartes' hands, the continuum between humans and the rest of the living world was sliced into a clear, unbridgeable dichotomy. This vision came to be known as *dualism*, and

Descartes' theory of matter came to be known as mechanical philosophy. It was an explicit attempt to disenchant the world – a direct attack on the remaining principles of animist philosophy. And from the 1630s, these ideas came to dominate science. We often think of the Church and science as antagonists, but in fact the architects of the Scientific Revolution were all deeply religious, and shared common cause with the clergy: to strip nature of spirit.

During the Enlightenment, dualist thought became mainstream for the first time in history. It gave sanction to the enclosure and privatisation of common land, as land was rendered but a thing to be possessed. And it was enclosure, in turn, that enabled dualism's rise to cultural dominance: only once commoners were alienated from the land and severed from forest ecosystems could they be convinced to imagine themselves as fundamentally separate from the rest of the living world, and to see other beings as objects.

Of course, the fallacies of mechanical philosophy couldn't last long. Within a century the notion of inert matter was debunked, as it became clear to scientists that animals and plants and other organisms are in fact alive.[34] But the damage was done. Dualism had taken hold in European culture. It became entrenched because it satisfied the need of powerful groups to divide the world in two. Once nature was an object, you could do more or less anything you wanted to it. Whatever ethical constraints remained against possession and extraction had been removed, much to the delight of capital. Land became property. Living beings became things. Ecosystems became resources.

Writing in the late 1700s, Immanuel Kant, one of Western philosophy's most celebrated ethicists, wrote: 'As far as non-humans are concerned, we have no direct duties. They are there merely as the means to an end. The end is man.'[35]

The body as 'raw material'

European elites leveraged Descartes' dualism to change people's beliefs about nature. But they also took it one step further, and sought to change people's beliefs about labour, too.

During the revolutionary period, peasant work followed a rhythm that from the perspective of industrialists appeared to be irregular and undisciplined: it depended on weather and seasons, on festivals and feast days. Life was organised around the principles of sufficiency and desire: people would work as much as they needed, and the rest of the time they spent dancing, telling stories, drinking beer . . . having *fun*. As the sociologist Juliet Schor puts it:

> The medieval calendar is filled with holidays . . . not only long 'vacations' at Christmas, Easter and midsummer but also numerous saints' and rest days. In addition to official celebrations, there were often weeks' worth of ales – to mark important life events (brides' ales or wake ales) as well as less momentous occasions (scot ale, lamb ale and hock ale). All told, holiday leisure time in England took up probably one-third of the year. And the English were apparently working harder than their neighbours. The *ancien regime* in France is reported to have guaranteed fifty-two Sundays, ninety rest days and thirty-eight holidays. In Spain, travellers noted that holidays totalled five months per year.[36]

According to the English historian E.P. Thompson, these festivals and carnivals 'were, in an important sense, what men and women lived for'.[37]

All of this posed a problem for the ruling class in the 1500s. Elites complained bitterly about the peasants' festivals, and castigated them for 'licentious behaviour and liberty'.[38] Peasant lifeways were incompatible with the kind of labour that was required for capital accumulation. Labour needed to go well beyond need; it needed to become a total way of life. Yes, enclosure helped solve this problem to some extent, by putting peasants at the mercy of hunger and forcing them to compete with each other. But it was not enough. In the wake of enclosure, Europe filled up with 'paupers' and 'vagabonds' – people who had been pushed off the land but either couldn't find work or otherwise refused to submit to the brutal conditions of the new capitalist farms and factories. They survived by begging, hawking and stealing food.

This problem preoccupied European governments for some three centuries. To deal with it, and assuage elite fears that the growing underclass might come to pose a political threat, states began to introduce laws forcing people to work. In 1531, England's King Henry VIII passed the first Vagabonds Act, describing 'idleness' as 'the mother and root of all vices' and ordering that vagabonds should be bound, whipped, and forced to 'put themselves to labour'. A series of other vagabond acts followed, each harsher than the one before. In 1547, Edward VI decreed that at the first offence vagabonds should be branded with a 'V' and subjected to two years of forced labour. The second offence was punishable by death.

These laws unleashed an extraordinary outpouring of state violence against the dispossessed. In England, as many as 72,000 'idle persons' were hanged during the reign of Henry VIII, according to one account. In the 1570s, up to 400 'rogues' were executed each year.[39] The goal was to fundamentally change people's beliefs about labour. Elites had to literally whip people into becoming docile, obedient, productive workers. During this time, philosophers and political theorists developed a peculiar fascination with the body, which they came to see as the repository of hidden labour-power, the key engine of capitalist surplus. The question was how to most efficiently extract the value that lay slumbering within.

Here too, Descartes came to the rescue. Dualism had established a clear divide between humans and nature, subject and object. But it was not only nature that was objectified in this new system. It was also the body. The body was recast as part of nature. In *Treatise of Man*, Descartes argued that humans are divided into two distinct components: an immaterial mind and a material body. The body – just like nature – was but brute matter, and its functions were like that of a machine. Descartes became enamoured of the anatomy theatre, where bodies were laid out in public and dissected, exposed as being mere flesh, profaned, devoid of spirit, composed of what amounted to ropes and pulleys and wheels. 'I am not my body,' Descartes insisted. Rather, it is disembodied thought, or mind, or reason, that constitutes the person. Thus the phrase by which we all know him: 'I think, therefore I am.'

Descartes succeeded in not only separating mind from body, but also establishing a hierarchical relationship between the two. Just as the ruling class should dominate nature and control it for the purposes of productivity, so the mind should dominate the body for the same purpose.

During the 1600s, Descartes' views were leveraged to bring the body under control, to defeat its passions and desires, and impose on it a regular, productive order. Any inclination towards joy, play, spontaneity – the pleasures of bodily experience – was regarded as potentially immoral. In the 1700s, these ideas coalesced into a system of explicit values: idleness is sin; productivity is virtue. In the Calvinist theology that was popular in Western Christianity at the time, *profit* became the sign of moral success – the proof of salvation. To maximise profit, people were encouraged to organise their lives around productivity.[40] Those who fell behind in the productivity race and slipped into poverty were branded with the stigma of sin. Poverty was recast not as the consequence of dispossession, but as the sign of personal moral failing.

These ethics of discipline and self-mastery became central to the culture of capitalism. The 'workhouses' that were built by parishes across Britain to absorb the 'idle' poor functioned partly as factories and partly as cultural re-education camps, rooting out any residual spirit of resistance while instilling the values of productivity, time and respect for authority. In the 1800s, factories developed timetables and the assembly line, with the purpose of squeezing maximum output from each worker. The early 1900s gave us Taylorism, with every tiny motion of a worker's body reduced to the most efficient possible movement. Work was progressively stripped of meaning, pleasure, talent and mastery.

There is nothing natural or innate about the productivist behaviours we associate with *homo economicus*. That creature is the product of five centuries of cultural re-programming.

Descartes' theory of the body made it possible to think of human labour as something that can be separated from the self, abstracted, and exchanged on the market – just like nature. As with land and nature, labour too was transformed into a mere commodity; a notion that would have been unthinkable only a century earlier. The refugees that enclosure was producing came to be seen not as subjects with rights, but as a mass of labour to be disciplined and controlled for the sake of capitalist growth.

Cheap nature

The 1600s gave rise to a new way of seeing nature: as something 'other', something separate from society – not just land, soils, forests and mountains, but also the bodies of human beings themselves. This new world view allowed capitalists to objectify nature and pull it into circuits of accumulation. But it also did something else. It allowed them to think of nature as 'external' to the economy. And because it was external it could be made cheap.

In order to generate profits for growth, capital seeks to appropriate nature as cheaply as possible – and ideally for free.[41] The elites' seizure of Europe's commons after 1500 can be seen as a massive, uncompensated appropriation of nature. So too with colonisation, when Europeans grabbed huge swathes of the global South; vastly more land and resources than Europe itself contained. Silver and gold from South America, land for cotton and sugar in the Caribbean, Indian forests for fuel and ship-building, and – during the scramble for Africa that got under way after 1885 – diamonds, rubber, cocoa, coffee, and countless other commodities. All of this was appropriated virtually for free. By 'free' here I mean not just in the sense that they didn't pay for it, but also in the sense that they gave nothing back. There was no gesture of reciprocity with the land. It was pure extraction; pure theft. In a system where nature is 'external', the costs of plundering it can be externalised.

Enclosure and colonisation enabled the appropriation of cheap labour too. And while capital paid wages, however meagre, to Europe's proletarian workers (mostly males), it did not pay for the (mostly female) labour that reproduced them: the women who cooked their food, cared for them when ill, and raised the next generation of workers. Indeed, it was enclosure that first produced the figure of the housewife that remains with us today, by cutting women off not only from the means of subsistence but from wage labour too, and confining them to reproductive roles. In the new capitalist system, a mass of hidden female labour was appropriated by elites virtually for free. Descartes' dualism was recruited for this task too. Within the dualist framework, bodies were set out on a spectrum. Women were regarded as closer to 'nature' than men. And they were treated accordingly – subordinated, controlled and exploited.[42] No need for compensation. As with everything shunted into the category of 'nature', the costs of extraction were externalised.

Something similar was playing out in the colonies, but there it was taken further still. During the colonial period, the peoples of the global South were routinely cast as 'nature': as 'savages', as 'wild', as less-than-human. Tellingly, the Spaniards referred to Indigenous Americans as *naturales*. Dualism was recruited in order to justify the appropriation not only of land in the colonies, but of the bodies of the colonised themselves. This played out clearly in the European slave trade. After all, in order to enslave someone, you first have to deny their humanity. Africans and Indigenous Americans were cast as objects in the European imagination, and exploited as such. As the Martiniquan writer Aimé Césaire put it, colonisation is, at base, a process of *thingification*.[43]

But there was also something else going on. The colonised were cast as 'primitive' precisely because they refused to accept the principles of human-nature dualism.[44] In the writings of European colonisers and missionaries we see they were dismayed that so many of the people they encountered insisted on seeing the world as alive – seeing mountains, rivers, animals, plants, and even the land as sentient beings with agency and spirit.

Europe's elites saw animist thought as an obstacle to capitalism – in the colonies just as in Europe itself – and sought to eradicate it. This was conducted in the name of 'civilisation.' To become civilised, to become fully human (and to become willing participants in the capitalist world economy), Indigenous people would have to be forced to abandon animist principles, and made to see nature as an object.

We all know that the violence of colonisation was justified, by its perpetrators, as part of a 'civilising mission'. What we tend not to grasp is that one of the key goals of this mission was to eradicate animist thought. The object was to *turn the colonised into dualists* – to colonise not only lands and bodies, but minds. As the Kenyan writer Ngũgĩ wa Thiong'o has put it: 'Colonialism imposed its control of the social production of wealth through military conquest and subsequent political dictatorship. But its most important area of domination was the mental universe of the colonised, the control, through culture, of how people perceived themselves and their relationship to the world.'[45]

Retweeting Descartes

We are all heirs of dualist ontology. We can see it everywhere in the language we use about nature today. We routinely describe the living world as 'natural resources', as 'raw materials', and even – as if to emphasise its subordination and servitude – as 'ecosystem services'. We talk about waste and pollution and climate change as 'externalities', because we believe that what happens to nature is fundamentally external to the concerns of humanity. These terms roll off our tongues and we don't even think twice about them. Dualism runs so deep that it wriggles into our language even when we're trying to be more conscientious. The very notion of 'the environment' – that thing we're supposed to care about – presupposes that the living world is nothing more than a passive container, a backdrop against which the human story plays out.

'Environment'. The strangeness of this innocent-seeming term becomes even clearer when we translate it into Spanish: *ambiente*. In the language of the conquistadors, the living world is cast as nothing more than mood lighting. From the perspective of animist ontology, this would be equivalent to regarding your mother and siblings as mere decorative portraits adorning the wall. It would be unthinkable.

These ideas didn't end with Bacon and Descartes. They have been retweeted and refined by a long parade of philosophers. Dualist assumptions show up even in postmodernist thought. Postmodernism prides itself on critiquing the hubris of Mind and Self and Truth, and on questioning grand metanarratives of human progress. And yet in the end all it does is take dualism to new extremes. The world, reality, doesn't really exist; or it does exist but it doesn't matter what it is, in itself, since reality is whatever humans construct it to be. Nothing really exists until it has been *realised* by humans, constituted in human language, given names and meaning, and inserted into our symbolic world. Reality outside our own experience literally dwindles into insignificance. Postmodernists may critique modernism, but only after accepting its basic terms.[46]

It's no wonder that we react so nonchalantly to the ever-mounting statistics about the crisis of mass extinction. We have a habit of taking this information with surprising calm. We don't weep. We don't get worked up. Why? Because we see humans as fundamentally separate from the rest of the living community. Those species are *out there*, in the *environment*. They aren't in here; they aren't part of *us*.[47] It is not surprising that we behave this way. After all, this is the core principle of capitalism: that the world is not really alive, and it is certainly not our kin, but rather just stuff to be extracted and discarded – and that includes most of the human beings living here too. From its very first principles, capitalism has set itself at war against life itself.

Descartes claimed that the purpose of science was 'to make ourselves the masters and possessors of nature'. Four hundred years later this ethic remains profoundly entrenched in our culture. We not only regard the living world as other, we regard it as an *enemy* – something that needs to be fought and subdued by the forces of science and reason. When Google executives created a new life sciences company in 2015, they named it 'Verily'. Asked to explain this odd name, Verily's CEO Andy Conrad said it had been chosen because 'only through the truth are we going to defeat Mother Nature'.

Two

Rise of the Juggernaut

> Capitalism can no more be 'persuaded' to limit growth than
> a human being can be 'persuaded' to stop breathing.
>
> Murray Bookchin

I still remember when I first learned about the history of capitalism in school. It was a happy story that started with the invention
of the steam engine in the eighteenth century and worked its
way through a parade of technological innovations, from the flying shuttle all the way up to the personal computer. I remember
marvelling at the glossy pictures in the textbook. As this story
would have it, economic growth is like a fountain of money that
springs forth from technology itself. It's a wonderful tale, and it
leaves us with the hopeful impression that with the right technology, we should be able to get growth more or less out of thin
air.

But when we think about the longer history of capitalism, it
becomes clear that something is missing from this story. Enclosure, colonisation, dispossession, the slave trade . . . historically,

growth has always been a process of appropriation: the appropriation of energy and work from nature and from (certain kinds of) human beings. Yes, capitalism has driven some extraordinary technological innovations, and these innovations have driven an extraordinary acceleration of growth. But the main contribution that technology makes to growth is not that it produces money out of thin air, but rather that it enables capital to expand and intensify the process of appropriation.[1]

This was true well before the steam engine. Even in the early 1500s, innovations in sugar-milling technology allowed plantation owners to put more land under sugar than they otherwise would have been able to process. Similarly, the invention of the cotton gin enabled producers to expand cotton monoculture. New wind-powered pumps were used to drain Europe's wild wetlands, opening vast tracts of land to farming. The development of bigger blast furnaces allowed for faster iron smelting, which in turn paved the way for more mining. And more logging was needed to fuel the furnaces, to the point where huge swathes of Europe's forests were felled to produce iron. The power of technology is that it enables capital and labour to be more productive – to produce more and faster. But it also speeds up the appropriation of nature.

In the nineteenth and twentieth centuries, this process was accelerated by the large-scale discovery of fossil fuel reserves – first coal and then oil – and the invention of technologies (like the steam engine) to extract and use them. A single barrel of crude oil can perform about 1700kWh of work. That's equivalent to 4.5 years of human labour. From the perspective of capital, tapping into underground oceans of oil was like colonising the Americas all over again, or a second Atlantic slave trade – a bonanza of appropriation. But it also supercharged the process of appropriation itself. Fossil fuels are used to power giant drills for deeper mining, trawlers for deep-sea fishing, tractors and combines for more intensive farming, chainsaws for faster logging, plus ships and trucks and aeroplanes to move all of these materials around the world at staggering speeds. Thanks to technology, the process of appropriation has become exponentially faster and more expansive.

We can see this acceleration reflected in the breathtaking speed at which GDP has shot up over the past century. But it would be a mistake to see this growth as *driven* by fossil fuels and technology. It has been *facilitated* by fossil fuels and technology, yes; but we have to ask ourselves: what is the deeper motivation, as it were, that propels capitalist growth?

The iron law of capital

A few months ago I found myself on stage for a televised debate about the future of capitalism, in front of a live audience. My opponent stood up and argued that there's nothing wrong with capitalism *as such*. The problem is that capitalism has been corrupted by greedy CEOs and venal politicians. All we need to do is deal with the bad apples and everything will be fine. After all, when it comes down to it, capitalism is just about people buying and selling things in the market – like your local farmers' market, or a souk in Morocco. These are innocent people using their skills to make a living; what could possibly be wrong with that?

It's a nice story, and it seems reasonable enough. But in fact the image here of small shops in farmers' markets and souks has nothing to do with capitalism. It is a false analogy. And it gets us no closer to understanding why capitalism is driving ecological breakdown. If we really want to understand how capitalism works, we need to dig a bit deeper.

The first step is to grasp that for most of human history, economies were organised around the principle of 'use-value'. A farmer might grow a pear because they like its juicy-sweet flavour or because it takes the edge off their hunger in the afternoon. An artisan might build a chair because it's useful for sitting on: to relax on the porch or to enjoy a meal around the table. And they might choose to sell these things in order to get money to buy other useful things, like a hoe for their garden or a pocketknife for their daughter. In fact, this is how most of us participate in the economy today. When we go to the shops it's usually to buy things that will be useful to us, like ingredients for dinner or a jacket to protect against the winter cold. We can summarise this kind of economy like this, where C stands for commodity (like a chair or a pear), and M stands for money:

$$C_1 \to M \to C_2$$

This might seem like a good description of capitalism on the face of it – free exchange of useful things between individuals. Just like in a farmers' market or a souk. But in reality there is nothing here that is particularly capitalist. It could be any economic system at all, at more or less any time or place in human history. What makes capitalism distinctive is that, for capitalists, value is reckoned quite differently. While a capitalist might recognise the usefulness of things like chairs and pears, the goal of producing them isn't to have a nice place to sit or a tasty afternoon snack, or even to sell them for other useful things. The goal is to produce and sell them for one purpose above all others: to make a profit. In this system, it is the 'exchange-value' of things that matters, not their use-value.[2] We can illustrate it like this, where the prime symbol (') represents an increase in quantity:

$$M \to C \to M'$$

This is the exact opposite of a use-value economy. But here's where things get interesting. Under capitalism, it's not enough to generate a steady profit. The goal is to reinvest that profit to expand the production process and generate yet more profit than the year before. We can illustrate it like this:

$$M \to C \to M' \to C' \to M'' \to C'' \to M''' \ldots$$

To understand what's going on here, we need to draw a distinction between two types of companies. Take your local restaurant, for example. It makes a profit at the end of the year, but the owners are content with more or less the same profit year after year: enough to pay the rent, put food on the table for their family, and maybe go for a holiday in the summer. While such a business might participate in elements of capitalist logic (paying wages, making a profit), it is not capitalist *as such*, since ultimately the profit is organised around some conception of use-value. This is how the vast majority of small businesses operate. Such shops existed thousands of years before capitalism emerged.

Now consider a corporation, like Exxon or Facebook or Amazon. A corporation doesn't operate according to the steady-state approach favoured by your local restaurant. Amazon's profits don't just go to putting food on the table for Jeff Bezos – they go into expanding the company: buying up competitors, putting local shops out of business, breaking into new countries, building more distribution centres, pumping out marketing campaigns to get people to buy stuff they don't need, all to extract more profit each year than the year before.

It's a self-reinforcing cycle – an ever-accelerating treadmill: money becomes profit becomes more money becomes more profit. And this is where we begin to see what makes capitalism distinctive. For capitalists, profit isn't just money at the end of the day, to be used for satisfying some specific need – profit becomes *capital*. And the whole point of capital is that it must be reinvested to produce more capital. This process never ends – it just continues expanding. Unlike your local restaurant, which is focused on satisfying particular concrete needs, there is no identifiable end point to the process of accumulating exchange-value. It is fundamentally unhinged from any conception of human need.

Looking at the formula above, it becomes clear that capital behaves a bit like a virus. A virus is a piece of genetic code that is programmed to replicate itself, but it cannot do so on its own: it has to infect a host cell and force that cell to create copies of its DNA, and then each of those copies goes on to infect other cells in order to create more copies, and so on. The sole purpose of a virus is self-replication. Capital too is built on a self-replicating code, and like a virus it seeks to turn everything it touches into a self-replicating replica of itself – more capital. The system becomes a juggernaut, an unstoppable machine that's programmed for endless expansion.

*

We often talk about the relentless expansionary drive of corporations like Amazon or Facebook as due to greed; CEOs like Mark Zuckerberg are just obsessed with accumulating money and power, we might say. But it's not quite so simple. The reality is that these firms, and the CEOs who run them, are subject to a *structural imperative* for growth. The Zuckerbergs of the world are just willing cogs in a bigger machine.

Here's how it works. Imagine you're an investor. You want returns of, say, 5% per year, so you decide to invest in Facebook. Remember, this is an exponential function. So if Facebook keeps churning out the same profits year after year (i.e., 0% growth), it will be able to repay your initial investment but it won't be able to pay you any interest on it. The only way to generate enough surplus for investor returns is to generate more profit each year than the year before. This is why when investors assess the 'health' of a firm, they don't look at net profits; they look at the *rate* of profit – in other words, how much the firm's profits grow each year. From the perspective of capital, profit alone doesn't count. It is meaningless. All that counts is growth.

Investors – people who hold accumulated capital – scour the globe in desperate search of anything that smells like growth. If Facebook's growth shows signs of slowing down, they'll pump their money into Exxon instead, or into tobacco companies, or into student loans – wherever the growth is at. This restless movement of capital puts companies under enormous pressure to do whatever they can to grow – in the case of Facebook, advertising more aggressively, creating ever-more addictive algorithms, selling users' data to unscrupulous agents, breaking privacy laws, generating political polarisation and even undermining democratic institutions – because if they fail to grow then investors will pull out and the firm will collapse. The choice is stark: grow or die. And this expansionary drive puts other companies under pressure, too. Suddenly no one can be satisfied with a steady-state approach; if you don't push to expand, you'll get gobbled up by your competitors. Growth becomes an iron law to which all are captive.

Why do investors engage in this restless quest for growth? Because when capital sits still, it loses value (due to inflation, depreciation, etc.). So as capital piles up in the hands of accumulators, it creates enormous pressures for growth. And the more that capital accumulates, the more the pressure builds.

Chasing the next fix

This becomes a problem because growth is a compound function. The global economy has typically grown at about 3% a year. This is what economists say is necessary to ensure that most capitalists accumulate profits. Three per cent doesn't sound like very much, but that's because our minds normally think of growth in linear terms. *Compound* growth – which is the basic structure of capital reinvestment – can be difficult to get our heads around. Indeed, it has an uncanny way of sneaking up on us.

There is an old fable that captures the surreal nature of growth – a tale about a mathematician in ancient India. To honour his achievements, the king summoned him to the palace and offered him a gift: 'Name whatever you want,' he said, 'and it is yours.'

The man responded humbly: 'My king, I am a modest man – all I ask is that you give me a bit of rice.' He took out a chessboard and continued: 'Put one grain on the first square, two on the second, four on the third, and continue doubling the grains on each square until you reach the end of the board. I will be content with that.'

The king thought it was an odd request but agreed, glad that the man had not asked for anything more extravagant.

By the end of the first row there were fewer than 200 grains on the board – not even enough for a meal. But then things became very strange. On the thirty-second square, only halfway through, the king had to place 2 billion grains – bankrupting his kingdom. If he had been able to continue, he would have had to place 18 million trillion grains on the sixty-fourth square, enough to cover the whole of India with rice a metre thick.

The same uncanny mechanism plays out when it comes to economic expansion. This tendency was noticed in 1772 by the mathematician Richard Price. Compound growth, he pointed out, 'increases at first slowly . . . but, the rate of increase being continuously accelerated, it becomes in some time so rapid as to mock all the powers of the imagination'.

Take the global economy in the year 2000 and grow it at the usual rate of 3% a year. Even at this modest-sounding increment, economic output will double every twenty-three years, which means *quadrupling* before the middle of the century, within half a human lifespan. And if we continue growing at that same rate, by the end of the century the economy will be twenty times bigger – twenty times more than we were already doing in the roaring 2000s. Another hundred years later and it's 370 times bigger. Another hundred years after that and it's 7,000 times bigger, and so on. It mocks all the powers of the imagination.

Some credit this aggressive energy for the rapid innovation that characterises capitalism. Certainly there is truth to that. But it also has the tendency to become extremely violent. Every time capital bumps up against barriers to accumulation (say a saturated market, a minimum-wage law, or environmental protections), then like a giant vampire squid it writhes in a desperate attempt to whip those barriers out of the way and plunge its tentacles into new sources of growth.[3] This is what is known as a 'fix'.[4] The enclosure movement was a fix. Colonisation was a fix. The Atlantic slave trade was a fix. The Opium Wars against China were a fix. The western expansion of the United States was a fix. Each one of these fixes – all of them violent – opened up new frontiers for appropriation and accumulation, all in service of capital's growth imperative.

In the nineteenth century the global economy was worth a little more than $1 trillion, in today's money. That means each year capital needed to find new investments worth about $30 billion – a significant sum. This required a huge effort on the part of capital, including the colonial expansion that characterised the nineteenth century. Today the global economy is worth over $80 trillion, so to maintain an acceptable rate of growth capital needs to find outlets for new investments worth another $2.5 trillion next year. That's the size of the entire British economy – one of the biggest in the world. Somehow we have to add the equivalent of another British economy next year, on top of what we are already doing, and then add even more than that the following year, and so on.

Where can this quantity of growth possibly be found? The pressures become enormous. It's what is driving the pharmaceutical companies behind the opioid crisis in the United States; the beef companies that are burning down the Amazon; the arms companies that lobby against gun control; the oil companies that bankroll climate denialism; and the retail firms that are invading our lives with ever-more sophisticated advertising techniques to get us to buy things we don't actually want. These are not 'bad apples' – they are obeying the iron law of capital.

Over the past 500 years, an entire infrastructure has been created to facilitate the expansion of capital: limited liability, corporate personhood, stock markets, shareholder value rules, fractional reserve banking, credit ratings – we live in a world that's increasingly organised around the imperatives of accumulation.

From private imperative to public obsession

But understanding the inner dynamics of capital only partly explains the growth imperative. To really grasp the pressures that are at play, we also have to pay attention to what governments are doing. Of course, governments have always been involved in advancing the interests of capitalist expansion. After all, enclosure and colonisation were ultimately backed up by the force of the state. But beginning in the early 1930s, during the Great Depression, something happened that added real fuel to these flames.

The Depression devastated the economies of the United States and Western Europe, and governments found themselves scrambling for a response. In the United States, officials reached out to Simon Kuznets, a young economist from Belarus, and asked him to develop an accounting system that would reveal the monetary value of all the goods and services that the US produced each year. The idea was that if you can see what is happening in the economy more clearly, you can figure out where things are going wrong and intervene more effectively. Kuznets created a metric called Gross National Product, which provided the basis for the Gross Domestic Product (GDP) metric we use today.

But Kuznets was careful to emphasise that GDP is flawed. It tallies up the market value of total production, but it doesn't care whether that production is helpful or harmful. GDP makes no distinction between $100 worth of tear gas and $100 worth of education. And, perhaps more importantly, it does not account for the ecological and social *costs* of production. If you cut down a forest for timber, GDP goes up. If you extend the working day and push back the retirement age, GDP goes up. If pollution causes hospital visits to rise, GDP goes up. But GDP says nothing about the loss of the forest as habitat for wildlife, or as a sink for emissions. It says nothing about the toll that too much work and pollution takes on people's bodies and minds. And not only does it leave out what is bad, it also leaves out much of what is good: it doesn't count most non-monetised economic activities, even when they are essential to human life and well-being. If you grow your own food, clean your own house or care for your ageing parents, GDP says nothing. It only counts if you pay companies to do these things for you.

Kuznets warned that we should never use GDP as a normal measure of economic progress. He thought we should improve it to account for the social costs of growth, so that governments would take human well-being into account and pursue more balanced objectives. But then the Second World War struck. As the Nazi threat mounted, Kuznets' concerns about well-being faded into the background. Governments needed to count *all* economic activities – even negative ones – so they could identify every shred of productive capacity and income available for the war effort. This more aggressive vision of GDP ended up becoming dominant. And at the Bretton Woods Conference in 1944, when world leaders sat down to decide the rules that would govern the world economy in the wake of the war, it was enshrined as the key indicator of economic progress – exactly what Kuznets had warned against.

Of course, there's nothing inherently wrong with measuring some things and not others. GDP itself doesn't have any impact in the real world, one way or the other. GDP *growth*, however, does. As soon as we start focusing on GDP growth, we're not only promoting the things GDP measures, we're promoting the indefinite increase of those things, regardless of the costs.

Initially, economists used GDP to measure 'levels' of economic output. Was the level too high, causing excess production and a glut of supply? Or was it too low, leaving people unable to get the goods they needed? During the Depression it was clear that output was too low – so to pull themselves out of it, Western governments invested heavily in infrastructure projects and created vast numbers of well-paid jobs, putting money into people's pockets to stimulate demand and get things moving again. It worked, and GDP went up. But growth was not a goal *in and of itself.* Remember, this was the progressive era of President Franklin Roosevelt. For the first time in history, the goal was to raise the level of output specifically in order to improve people's livelihoods and achieve progressive social outcomes – quite unlike during the previous 400 years. In other words, early progressive governments treated growth as a use-value.

But that didn't last long. When the OECD was founded in 1960, the top goal in its charter was (and remains) to 'promote policies designed to achieve the highest sustainable rate of economic growth'. Suddenly the objective was to pursue not just higher levels of output for some specific purpose, but *the highest* rate, indefinitely, for its own sake. The British government followed suit, setting a target of 50% growth over the course of a single decade – an extraordinary rate of expansion, and the first time growth for its own sake was enshrined as a national policy objective.[5]

The idea spread like wildfire. During the Cold War, the grand competition between the West and the USSR came to be adjudicated largely by rates of growth. Which system could grow GDP the fastest? And of course growth was not only symbolically powerful in this contest; to the extent that it enabled more investment in military capacity, it also translated into real geopolitical advantage.

This new focus on GDP growth for its own sake – growthism – forever changed the way that Western governments managed their economies. The progressive policies that had been used to improve social outcomes after the Great Depression, like higher wages, labour unions and investment in public health and education, suddenly became suspect. These policies had led to high levels of well-being, but in so doing had made labour too 'expensive' for capital to maintain a high rate of profit. So too with the environmental regulations that were rolled out during this period, which restricted the exploitation of nature (the US Environmental Protection Agency was founded in 1970). In the late 1970s, growth in Western economies began to slow down and returns on capital began to decline. Governments came under pressure to do something about it – to create a 'fix' for capital. So they attacked unions and gutted labour laws in order to drive the cost of wages down, they dismantled key environmental protections, and they privatised public assets that had previously been off limits to capital – mines, railways, energy, water, healthcare, telecommunications and so on – creating lucrative opportunities for private investors. During the 1980s this strategy was pursued with particular zeal by Ronald Reagan in the US and Margaret Thatcher in the UK, inaugurating the approach that today we call neoliberalism.[6]

Some people have a tendency to think of neoliberalism as a mistake – an overly-extreme version of capitalism that we should reject in favour of returning to the somewhat more humane version that prevailed in prior decades. But the shift to neoliberalism was not a mistake; it was driven by the growth imperative. In order to restore the rate of profit and keep capitalism afloat, governments had to shift away from social objectives (use-values) to focus instead on improving the conditions for capital accumulation (exchange-value). The interests of capital came to be internalised by the state, to the point where today the distinction between growth and capital accumulation has almost completely collapsed. Now the goal is to tear down barriers to profit – to make humans and nature cheaper – for the sake of growth.

Western governments also pushed this agenda across the global South, as part of the same fix: to open up new frontiers for capital. After the end of colonialism in the 1950s, many newly independent governments had been developing a new direction in economics. They were rolling out progressive policies to rebuild their countries, using tariffs and subsidies to protect domestic industries; improving labour standards and raising workers' wages; and investing in public healthcare and education. All of this was intended to reverse the extractive policies of colonialism and improve human welfare – and it was working. Average income in the global South grew at 3.2% per year during the 1960s and 1970s. Crucially, in most cases growth was not pursued as a goal in and of itself; it was a means to recovery, independence, and human development – much as it was for the West in the years after the Great Depression.

But Western powers were not happy with this turn of events, as it meant they were losing access to the cheap labour, raw materials and captive markets that they had enjoyed under colonialism. So they intervened. During the debt crisis of the 1980s, they leveraged their power as creditors and used their control over the World Bank and the International Monetary Fund (IMF) to impose 'structural adjustment programmes' across Latin America, Africa and parts of Asia (with the exception of China and a few other East Asian countries). Structural adjustment forcibly liberalised the economies of the global South, tearing down protective tariffs and capital controls, cutting wages and environmental laws, slashing social spending and privatising public goods – all to break open profitable new frontiers for foreign capital and restore access to cheap labour and resources.[7]

Structural adjustment fundamentally reshaped the economies of the South. Governments were forced to abandon their focus on human welfare and economic independence and focus instead on creating the best possible conditions for capital accumulation. This was done in the name of growth, but the consequences for the South were disastrous. The imposition of neoliberal policies caused two decades of crisis, with rising poverty, inequality and unemployment. Income growth rates across the South collapsed during the 1980s and 1990s, down to an average of 0.7% over this two-decade period.[8] But as far as capital was concerned, it worked like a charm: it enabled multinational companies to post record profits, and sent the incomes of the richest 1% soaring.[9] Western growth rates recovered, which was the real objective of structural adjustment (it was a fix!), but at the expense of human lives across the South. The legacy of this intervention has been an extraordinary increase in global inequality over the past few decades. The real per capita income gap between the global North and global South is four times larger today than it was at the end of colonialism.[10]

The straight-jacket

Today, nearly every government in the world, rich and poor alike, is focused single-mindedly on GDP growth. This is no longer a matter of choice. In a globalised world where capital can move freely across borders at the click of a mouse, nations are forced to compete with one another to attract foreign investment. Governments find themselves under pressure to cut workers' rights, slash environmental protections, open up public land to developers, privatise public services – whatever it takes to please the barons of international capital in what has become a global rush towards self-imposed structural adjustment.[11] All of this is done in the name of growth.

The governments of the world are bound to a new rule: not to achieve a level of output adequate to improve wages and build social services, but rather to pursue growth *for its own sake*. The concrete use-values of economic production (meeting human needs) have been subordinated to the pursuit of abstract exchange-value (GDP growth). Governments justify this by saying that GDP growth is the only way to reduce poverty, to create jobs and to improve people's lives. Indeed, growth has come to stand in for human well-being, and even progress itself. This is remarkable, given that GDP measures such a narrow slice of economic activity. GDP growth is, ultimately, an indicator of the welfare of capitalism. That we have all come to see it as a proxy for the welfare of humans represents an extraordinary ideological coup.

Of course, in some respects it's true. In capitalist economies, people's livelihoods are tied to GDP growth. We all need jobs and wages in order to survive. And here's where the problems begin. Under capitalism, companies are constantly finding ways to increase labour productivity in order to push down the costs of production. As labour productivity improves, firms need fewer workers. People get laid off and unemployment rises; poverty and homelessness go up. Governments have to respond by scrambling to generate more growth just to create new jobs. But the crisis never goes away; it just keeps recurring, year after year. This is known as the 'productivity trap'.[12] We are in the absurd position of needing perpetual growth just in order to avoid societal collapse.

There are other traps governments find themselves in. If a government wants to invest in public healthcare and education, it has to find (or create) the money to do so. One option is to raise taxes on the rich and on corporations, but in countries where moneyed interests have political influence this risks triggering a backlash. Given this risk, even progressive parties find themselves on the horns of a dilemma. How do you get the resources to improve the lives of ordinary people without turning powerful rich people against you? Growth.

Then there's the debt trap – one of the most powerful of the growth imperatives. Governments finance their activities in large part by selling bonds, which is a way of borrowing money. But bonds come with interest, and interest is a compound function. In order to pay interest on bonds, governments have to generate revenues, which usually means pursuing growth. When economies slow down, governments can't pay their debts, triggering a crisis that can quickly spiral out of control: bonds lose their value, and in order to sell them governments have to promise higher interest rates, putting them yet further into debt. The only way to get out of such a crisis is to start slashing any 'barriers' to growth – labour laws, environmental protections, capital controls, anything to give investors the 'confidence' they need to keep buying bonds. Just like companies, governments face a stark choice: grow the economy or collapse.

On top of all this, governments pursue growth because GDP is the currency of international political power. This is clearest in military terms: the bigger your GDP, the more tanks, missiles, aircraft carriers and nuclear weapons you can buy. But it's also true in economic terms. For example, a nation's bargaining power at the World Trade Organization depends on the size of its GDP. The biggest economies are able to push through trade deals that serve their own interests, and they're able to wield sanctions as a weapon to force smaller economies to fall in line. Governments find themselves scrambling in a desperate, dog-eat-dog competition to get to the top of the pile, just to avoid being pushed around. Geopolitical pressure has become a powerful driver of the growth imperative.

Growth is so deeply embedded in our economics and politics that the system can't survive without it. If growth stops, companies go bust, governments struggle to fund social services, people lose their jobs, poverty rises, and states become politically vulnerable. Under capitalism, growth is not just an optional feature of human social organisation – it's an *imperative* to which all are hostage. If the economy doesn't grow, everything falls apart. We're in a straight-jacket. So it's no surprise that governments around the world have placed the full force of the state behind perpetuating the treadmill of accumulation.

All of this has powered an extraordinary acceleration of GDP since 1945. And from the perspective of ecology, this is where things start to go wrong.

A world devoured

None of this is to say that growth is bad, in and of itself. That's not my argument. It's not growth that's the problem, it's *growthism*: the pursuit of growth for its own sake, or for the sake of capital accumulation, rather than to meet concrete human needs and social objectives. When we look at the impact that growthism has had on our planet since the 1980s, it makes the period of enclosure and colonisation seem quaint by comparison. All of the land and resources that colonisers appropriated across multiple continents and pulled into the juggernaut of capital – all of that has been dwarfed many times over.

We can see this playing out in the statistics on raw material consumption. This metric tallies up the total weight of all the stuff humans extract and consume each year, including biomass, metals, minerals, fossil fuels and construction materials. These figures tell an astonishing story. They show a steady rise of material use in the first half of the 1900s, doubling from 7 billion tons per year to 14 billion tons per year. But then, in the decades after 1945, something truly bewildering happens. As GDP growth becomes entrenched as a core political objective around the world, and as economic expansion starts to accelerate, material use explodes: it reaches 35 billion tons by 1980, hits 50 billion tons by 2000, and then screams up to an eye-watering 92 billion tons by 2017.[13]

The graph on page 102 is almost breathtaking to look at. Of course, some of this increase represents important improvements in people's access to necessary goods (in other words, use-value), particularly in poorer parts of the world; and we should celebrate that. But most of it does not. Scientists estimate

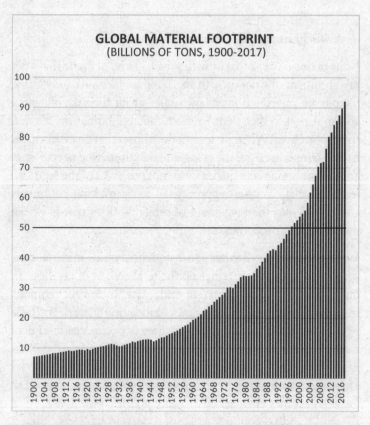

GLOBAL MATERIAL FOOTPRINT
(BILLIONS OF TONS, 1900-2017)

The horizontal black line indicates what scientists consider to be the maximum sustainable threshold (Bringezu 2015). Source: Krausmann et al. (2009), materialflows.net

that the planet can handle a total material footprint of up to about 50 billion tons per year.[14] That's considered to be a maximum safe boundary. Today we're exceeding that boundary twice over. And, as we will see, virtually all of this overshoot is being driven by excess consumption in high-income nations – consumption that is organised not around use-value but exchange-value.

Keep in mind that every ton of material stuff that's extracted from the earth comes with an impact on the planet's living systems. Ramping up the extraction of biomass means razing forests and draining wetlands. It means destroying habitats and carbon sinks. It means soil depletion, ocean dead zones and overfishing. Ramping up the extraction of fossil fuels means more carbon emissions, more climate breakdown and more ocean acidification. It means more mountaintop removal, more offshore drilling, more fracking and more tar sands. Ramping up the extraction of ores and construction materials means more open-cast mining, with all the downstream pollution that entails, and more cars and ships and buildings that demand yet more energy. And all this entails more waste: more landfills in the countryside, more toxins in our rivers, and more plastics in the sea. According to the United Nations, material extraction and processing is responsible for 90% of total global biodiversity loss.[15] In fact, scientists often use material footprint as a proxy for ecological damage itself.[16]

The rise in material use after 1945 reflects what scientists have called the Great Acceleration – the most aggressive and destructive period of the Capitalocene. Virtually every indicator of ecological impact has exploded as a result.

This increase in material use tracks more or less exactly with the rise of global GDP. The two have grown together in lockstep. Every additional unit of GDP means roughly an additional unit of material extraction. There were times, such as during the 1990s, when GDP grew at a slightly faster rate than material use, prompting some to hope we were on our way to decoupling GDP from material use altogether. But those hopes have been dashed in the decades since. In fact, exactly the opposite has happened. Since 2000, the growth

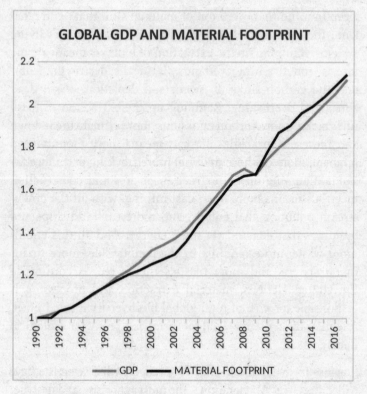

GLOBAL GDP AND MATERIAL FOOTPRINT

Source: materialflows.net, World Bank

of material use has *outpaced* the growth of GDP. Instead of gradually dematerialising, the global economy has been *re*materialising.

Perhaps most disturbingly of all, this trend shows no signs of slowing down. On our present trajectory, with business as usual, we are on course to be using more than 200 billion tons of material stuff per year by the middle of the century, more than double what we're using right now. That's four times over the safe boundary. There's no telling what kind of ecological tipping points we might trip in the process.

*

We can see exactly the same thing happening when it comes to climate change. We normally think of climate change as being driven by emissions from fossil fuels. And of course this is true. But there is a deeper mechanism at play that we too often ignore. Why are we burning through so much fossil fuel in the first place? Because economic growth requires energy. For the entire history of capitalism, growth has always caused energy use to rise.[17]

This is hardly surprising. After all, it requires an extraordinary amount of energy to extract and process and transport all the material stuff the global economy devours each year. There has been a radical acceleration of fossil fuel use since 1945, rising along with both GDP and material use. And carbon emissions have gone up right along with it. Annual emissions more than doubled from 2 billion tons per year to 5 billion tons per year during the first half of the 1900s. During the second half of the century they rose fivefold, reaching 25 billion tons by the year 2000. And they have continued to rise since then, despite a string of international climate summits, reaching 37 billion tons in 2019.

Of course, there is no *intrinsic* relationship between energy use and CO_2 emissions. It all depends on what energy source we're using. Coal is by far the most carbon-intensive of the fossil fuels. Oil – which has grown much more quickly than coal since 1945 – emits less CO_2 per unit of energy. And natural gas is less intensive still.[18] As the global economy has come to rely more on these less polluting fuels, one might think that emissions would begin to decline. This has happened in a number of high-income nations, but not on a global scale. Why? Because GDP growth is driving total energy demand up at such a rapid pace that these new fuels aren't *replacing* the older ones, they are being added on top of them. The shift to oil and gas hasn't been an energy transition, but an energy *addition*.

The same thing is happening right now with renewable energy. Over the past couple of decades there has been extraordinary growth in renewable energy capacity, which is worth celebrating. In some nations, renewables have begun to displace fossil fuels. But on a global scale, growth in energy demand is swamping growth in renewable capacity. All that new clean energy isn't replacing dirty energies, it's being added on top of them.[19] This dynamic should give us pause. Yes, we need as much renewable energy as we can get – but it won't make enough of a difference if the global economy continues to grow at existing rates. The more we grow, the more energy the global economy requires, the more difficult it is to cover it with cleaner energy sources.

*

All of this changes how we think of GDP growth. We have been trained to see exponentially rising GDP as a proxy for human progress. But it's not quite so simple. We need to retrain our eyes. It's like looking at one of those pictures that seems like an ordinary two-dimensional pattern, but when you change your focus and look more deeply then suddenly a new three-dimensional image comes into view. A more holistic way of thinking about growth is to recognise that it is broadly equivalent to the rate at which our economy is metabolising the living world. This is not a problem, in and of itself; but past a certain point – which, as we will see, rich nations have long since surpassed – it becomes extremely destructive. Under capitalism, the rate of growth is the rate at which nature and human lives are being commodified and roped into circuits of accumulation. That we have come to rely on this as our primary indicator of progress reveals the extent to which we have come to see the world from the perspective of capital rather than from the perspective of life. Indeed, there is a bitter irony to the fact that we have been persuaded to use the word 'growth' to describe what has now become primarily a process of breakdown.

Colonialism 2.0

But there's something wrong with this picture. The language I've been using here – the language of 'we' – isn't quite accurate. Even when we accept that capitalism is driving ecological breakdown, we have a tendency to describe it in collective terms, as if all humans are equally responsible. The ideology of the Anthropocene has a way of worming its way back into our discourse. But this assumption blinds us to what's really going on. The word 'Anthropocene' is wrong not just because previous economic systems did not pose a threat to global ecology in the way that capitalism does today. It's also that even today not all people are equally responsible.

Once we grasp the relationship between GDP growth and ecological impact, it's easy enough to guess that countries with higher GDP per capita will have higher ecological impact, and vice versa. And that's exactly how it plays out. We can see this disparity in virtually every category of consumption for which we have data. Take meat, for example, which we know has a significant ecological footprint. In India, the average person consumes 4 kilograms of meat per year. In Kenya they consume 17 kilograms. In the United States, it's a staggering 120 kilograms. The average US American consumes more meat each year than thirty Indians.[20] Or look at plastic – another major ecological hazard. In the Middle East and Africa, the average person gets through 16 kilograms of plastic per year. That's a lot. But in Western Europe it's nine times higher: 136 kilograms per person per year.[21]

We can see the same pattern playing out when it comes to material footprint. Low-income countries consume only about 2 tons of material stuff per person per year. Lower-middle-income countries consume about 4 tons per person, and upper-middle-income countries consume about 12. As for high-income nations, they consume many times more than this: about 28 tons per person per year, on average. In the United States, it's 35 tons. To put this in perspective, ecologists say that a sustainable level of material footprint, rendered in per capita terms, is about 8 tons per person. High-income nations blow past that boundary nearly four times over.[22]

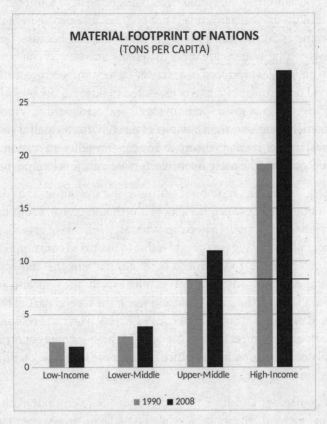

The horizontal black line indicates the sustainable threshold in per capita terms (cf. Bringezu 2015).[23] Source: materialflows.net

It doesn't take a mathematician to calculate who's responsible for the mess we're in. Consider this: if high-income nations were to consume at the average level of the rest of the world, we would not be overshooting the safe boundary at all. We'd be operating roughly within the planet's biocapacity, rather than staring down the barrel of an ecological emergency. By contrast, if everyone in the world were to consume at the level of high-income countries, we would need the equivalent of four planets to sustain us. Crucially, this is not only because people in high-income countries consume more stuff; it's also because their provisioning systems are more materially intensive. If you buy a can of Pringles produced in a faraway factory, shipped across the world on aeroplanes and trucks, stored in huge warehouses and packaged in copious amounts of plastic and cardboard, it is more materially intensive than buying potato chips from a stall at your local farmers' market. The more an economy relies on corporate supply chains, the more intensive its material use is likely to be.

These inequalities have been getting worse over time. The consumption gap between the global North and global South has exploded since 1990. In per-capita terms, a full 81% of growth in material use during this period is due to increased consumption in rich nations. If we want to build a more humane and ecological economy, we need to be doing exactly the opposite: we need to shrink the gap. As we will see in the second part of this book, most global South countries will need to increase resource use in order to meet human needs, while high-income countries will need to dramatically reduce resource use to get back within sustainable levels.

Of course, we also have to think about the role of population going forward. The more the global population grows, the more difficult this challenge will be. As we approach this question, it's crucial – as always – that we focus on underlying structural drivers. Many women around the world do not have control over their bodies and the number of children they have. Even in liberal nations women come under heavy social pressure to reproduce, often to the point where those who choose to have fewer or no children are interrogated and stigmatised. Poverty exacerbates these problems considerably. And of course capitalism itself creates pressures for population growth: more people means more labour, cheaper labour, and more consumers. These pressures filter into our culture, and even into national policy: countries like France and Japan are offering incentives to get women to have more children, to keep their economies growing.

It's essential that we stabilise the size of the human population. The good news is that this is not a particularly difficult matter: as the economist Kate Raworth put it to me, 'It's one growth curve that the world actually knows how to flatten, so it's not the one that keeps me awake at night.' What brings a nation's birth rate down? Investing in child health, so that parents can be confident their children will survive; investing in women's health and reproductive rights, so that women have greater control over their own bodies and family size; investing in girls' education to expand their choices and opportunities; and ensuring economic security for all. With these policies in place, population growth falls fast – even within a single generation.[24] Gender justice and economic justice must be central to any vision for a more ecological economy.

But stabilising the global population would not cause ecological damage to automatically level off, in and of itself. In the absence of *more* consumers, capital finds ways to get *existing* consumers to consume more. Indeed, that has been the dominant story for the past few hundred years: the growth rate of material use has always significantly outstripped the growth rate of the population. Indeed, material use keeps rising even when populations stabilise and decline. This has been the case in most historical examples of population stability under capitalism.

The data on material consumption shows that high-income countries are the biggest drivers of ecological breakdown. But there's another side to this equation: we also have to ask where in the world that breakdown is happening. High-income nations depend in large part on extraction from the global South. In fact, fully half of the total materials they consume are extracted from poorer countries, and generally under unequal and exploitative conditions. The coltan in your smartphone comes from mines in the Congo. The lithium in your electric car battery comes from the mountains of Bolivia. The cotton in your bedsheets comes from plantations in Egypt. And this dependency does not run in the other direction. The vast majority of materials that are consumed in the South ultimately originate from the South itself, even if they are cycled through multinational value chains.[25]

In other words, there is an enormous *net* flow of resources that goes from poor countries to rich countries, including around 10 billion tons of raw materials per year. The patterns of extraction that characterised colonisation remain very much in place today. But this time, instead of being seized by force, those resources are being extracted and sold, for cheap, by governments that have been rendered dependent on foreign investment and beholden to the growth imperatives of capitalism.

*

We can see similar patterns of inequality playing out when it comes to climate breakdown. You wouldn't know it from the dominant narrative, though. The media tend to focus on each country's current territorial emissions. By this metric, China is the biggest culprit by far. China emits 10.3 gigatons of CO_2 per year, almost double that of the United States, which comes in as the second worst offender. The European Union is third, but India is not far behind, and emits more than major industrial nations like Russia and Japan.

Looking at the data from this angle we might be tempted to conclude that responsibility for the climate crisis is shared more or less equally between the nations of the global North and the global South. But there are a number of problems with this approach. First, it doesn't correct for population size. When we look at it in *per capita* terms, the story changes completely. India emits only 1.9 tons of CO_2 per person. In China it's 8 tons per person. By contrast, Americans emit more than 16 tons per person – double that of China and eight times more than Indians. Plus, we also have to account for the fact that, since the 1980s, high-income nations have outsourced much of their industrial production to poorer countries in the global South, to take advantage of cheap labour and resources, thereby shifting a big chunk of their emissions off the books. If we want a more accurate picture of national responsibility, we need to look beyond just territorial emissions and count consumption-based emissions too.

But the biggest problem with the usual media narrative is that when it comes to climate breakdown, what matters is the *stocks* of carbon dioxide in the atmosphere, not annual flows. So we need to look at each country's *historical* emissions. When we approach it this way, it becomes clear that the highly industrialised nations of the global North – in particular the United States and Western Europe – are responsible for the vast majority of the problem.

One way to take all this into account is to start from the principle of 'atmospheric commons': recognising that the atmosphere

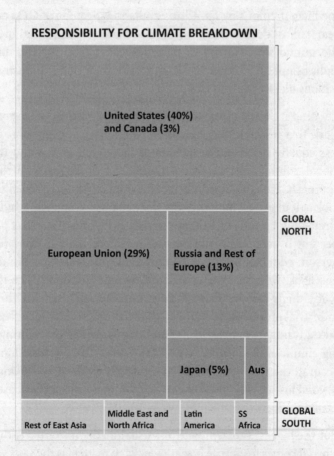

This image depicts historical emissions in excess of national fair shares of the 350ppm boundary (territorial emissions from 1850–1969, consumption-based emissions from 1970–2015). Source: Hickel 2020. Data management by Huzaifa Zoomkawala.[26]

is a finite resource, and all people are entitled to an equal share of it within the safe planetary boundary, which scientists have defined as an atmospheric CO_2 concentration of 350 parts per million. Using this framework, we can measure the extent to which nations have exceeded or 'overshot' their safe fair share, and thus how much they have contributed to climate breakdown. The graph on page 114 depicts the results, counting historical emissions from 1850 to 2015, and using consumption-based emissions wherever possible.

The figures are staggering. The United States is single-handedly responsible for no less than 40% of global overshoot emissions. The European Union is responsible for 29%. Together with the rest of Europe, plus Canada, Japan and Australia, the nations of the global North (which represent only 19% of the global population) have contributed 92% of overshoot emissions. That means they are responsible for 92% of the damage caused by climate breakdown. By contrast, the entire continents of Latin America, Africa and the Middle East have contributed a combined total of only 8%. And that comes from only a small number of countries within those regions.[27]

In fact, the vast majority of global South countries have emitted so little in historical terms that they are still *under* their fair share of the planetary boundary. India is still 90 gigatons under its fair share. Nigeria is under by 11 gigatons, and Indonesia by 14 gigatons. Even China was under its fair share in 2015, by a full 29 gigatons, although given the sheer scale of China's present emissions it has probably exceeded this budget in the years since. In other words, the higher-income countries that have gobbled up not only their own fair shares but also everybody else's owe a climate debt to the rest of the world.

What's happening here should be understood as a process of atmospheric colonisation. A small number of high-income nations have appropriated the vast majority of the safe atmospheric commons, and have contributed the vast majority of emissions in excess of the planetary boundary.

This process of atmospheric colonisation is not unrelated to the earlier process of colonisation proper. We know that the North's industrial rise was enabled by the colonial appropriation of land, resources and bodies from the South. The data we now have on historical emissions reveals that the North's industrialisation has also been a process of appropriating the atmosphere – what we might call atmospheric theft. And just as the first phase of colonisation wrought ecological and human destruction across the South, now so too is this. Ironically, despite having contributed virtually nothing to the climate crisis, the South bears the vast majority of the impact of climate breakdown.

We are all aware of the climate damages the global North suffers. The hurricanes that strike the United States, the floods that swamp the UK each winter, the heatwaves that scorch Europe and the brutal fires that have ravaged Australia. These devastating stories dominate our headlines, and journalists are right to cover them. But they pale in comparison to the disasters that have been inflicted on the South – stories which appear only fleetingly on our screens, when they appear at all, like the storms that have decimated so much of the Caribbean and Southeast Asia, and the droughts in Central America, East Africa and the Middle East that have pushed people into hunger and forced them to flee their homes. Comparatively speaking, North America, Europe and Australia are among the least vulnerable to the effects of climate change. The real damage is happening across Africa, Asia and Latin America – and it's happening on a truly dystopian scale.

One way of illustrating these inequalities is to look at the distribution of monetary costs. According to data from the Climate Vulnerability Monitor, the South bears 82% of the total costs of climate breakdown, which in 2010 added up to $571 billion in losses due to drought, floods, landslides, storms and wildfires.[28]

Researchers predict that these costs will continue to rise. By 2030 the South will suffer 92% of total global costs, reaching $954 billion.

The distribution of climate-change-related deaths is even more skewed towards the South. Data from 2010 indicates that around 400,000 people died that year due to crises related to climate breakdown – mostly hunger and communicable disease. No fewer than 98% of these deaths occurred in the South. And the vast majority, 83%, occurred in the countries that have the lowest carbon emissions in the world. By 2030, climate-related deaths are projected to reach up to 530,000 a year. Virtually all of these will happen in the South. Rich countries will suffer only 1% of climate-related deaths within their borders.

Why are the impacts of climate change so unevenly distributed? For one thing, climate change is causing rainfall patterns to shift north. As a result, drought-prone areas of the global South will have even less water than they do now. This will have devastating consequences for the region's agriculture, where crop yields are predicted to decline faster than the world average. Disease is another important factor. Rising temperatures are expanding the range of tropical diseases like malaria, meningitis, dengue and zika. But it's also because communities in the global South are, after a long history of colonisation and structural adjustment, least able to adapt to climate breakdown. This is particularly true for the poorest, who are more likely to live on marginal land vulnerable to droughts and floods, who don't have the financial buffer to see themselves through disasters, and who cannot easily relocate, or find new livelihoods, or defend their human rights. That the excess emissions of a few rich nations will harm billions of people in poorer nations is a crime against humanity and we should have the clarity to call it that. As Philip Alston, the UN Special Rapporteur on extreme poverty and human rights, has put it: 'Climate change is, among other things, an unconscionable assault on the poor.'[29]

This assault is already happening. Take Somaliland, for example – a small nation in the Horn of Africa. Over the past few years, a series of consecutive droughts has killed 70% of the country's livestock, devastating rural communities and forcing tens of thousands of families to flee. 'We used to have droughts before,' said Shukri Ismail Bandare, the Minister for Environment, in an interview with the *Financial Times*. 'We used to name the droughts. They would be 10 or 15 years apart. Now it is so frequent that people cannot cope with it. You can touch it in Somaliland, the climate change – it is real, it is here'.[30]

Remember, this is happening at 1°C. Two degrees will be a death sentence for much of the global South. The only reason that people have come to accept 2°C as a reasonable target is because climate negotiators from the United States and other powerful countries have pushed for it, over the loud objections of their colleagues from the South – and particularly from Africa. When the 2°C target was announced at the Copenhagen summit in 2009, Lumumba Di-Aping, the Sudanese chief negotiator for the G77, said: 'We have been asked to sign a suicide pact.' 'It is unfortunate,' he went on, 'that after 500 years-plus of interaction with the West we are still considered "disposables".' Cheap nature, he might have added.

The trauma of climate breakdown in the South directly echoes the trauma of colonisation. The South has suffered twice over: first from the appropriation of resources and labour that fuelled the North's industrial rise, and now from the appropriation of atmospheric commons by the North's industrial emissions. If our analysis of the climate crisis is not attentive to these colonial dimensions, then we have missed the point.

How to think about 'limits' in the 21st century

The thing about growth is that it sounds so *good*. It's a powerful metaphor that's rooted deeply in our understanding of natural processes: children grow, crops grow . . . and so too the economy should grow. But this framing plays on a false analogy. The natural process of growth is always finite. We want our children to grow, but not to the point of becoming 9 feet tall, and we certainly don't want them to grow on an endless exponential curve; rather, we want them to grow to a point of maturity, and then to maintain a healthy balance. We want our crops to grow, but only until they are ripe, at which point we harvest them and plant afresh. This is how growth works in the living world. It levels off.

The capitalist economy looks nothing like this. Under capital's growth imperative, there is no horizon – no future point at which economists and politicians say we will have enough money or enough stuff. There is no *end*, in the double sense of the term: no maturity and no purpose. The unquestioned assumption is that growth can and should carry on for ever, for its own sake. It is astonishing, when you think about it, that the dominant belief in economics holds that no matter how rich a country has become, their GDP should keep rising, year after year, with no identifiable end point. It is the definition of absurdity. We do see this pattern playing out in nature, sometimes, but only with devastating consequences: cancer cells are programmed to replicate for the sake of replicating, but the result is deadly to living systems.

To imagine that we can continue expanding the global economy indefinitely is to disavow the most obvious truths about our planet's ecological limits. This realisation first struck home in

1972, when a team of scientists at MIT published a groundbreaking report titled *Limits to Growth*. The report outlined findings from the team's cutting-edge work using a powerful computer model called World3, which was designed to analyse complex ecological, social and economic data from 1900 to 1970, and to predict what would happen to our world in twelve different scenarios by the end of the twenty-first century.

The results were striking. The business-as-usual scenario, with economic growth continuing at its normal rate, showed that sometime between 2030 and 2040 we would run into a crisis. Driven by the compound nature of the growth function, renewable resources would begin to reach the limits of their renewability, non-renewable resources would begin running out, and pollution would begin to exceed the capacity of the Earth to absorb it. Nations would have to spend increasing amounts of money to try to solve these problems, thereby spending less on the reinvestment that's required to keep generating growth. Economic output would begin to fall, the food supply would stagnate, living standards would diminish, and populations would begin to shrink. 'The most probable result,' they wrote, somewhat ominously, 'will be a rather sudden and uncontrollable decline in both population and industrial capacity.'

It touched a nerve. *Limits to Growth* exploded onto the scene and became one of the best-selling environmental titles in history, tapping into the countercultural ethos that prevailed in the wake of the youth rebellions of 1968.

But then the backlash came – and it came with overwhelming force. The report was denounced in the pages of the *Economist*, *Foreign Affairs*, *Forbes* and the *New York Times*, and big-name economists came out railing against it. They said that the model was too simplistic. It didn't account for the seemingly limitless innovation of which capitalism is capable. Sure, existing reserves of non-renewable resources might run out, but new technologies would enable us to find new reserves, or ways to use substitute materials. And yes, there might be limits to the amount of land available for renewable resources like food, but we can always develop better fertilisers and more productive crop varieties, or grow food in warehouses.

The Oxford professor Wilfred Beckerman went so far as to say that, thanks to the wonders of technological progress, there is 'no reason to suppose that economic growth cannot continue for another 2,500 years'. Ronald Reagan ran an election campaign against incumbent President Jimmy Carter – an environmentalist – by attacking the notion of limits, and linking a celebration of limitlessness to the spirit of the American Dream itself. 'There is no such thing as limits to growth,' he said, 'because there is no such thing as limits to the human imagination.' It was a winning message, and Americans bought it. Reagan beat Carter in a landslide.

During the decade that followed, with the collapse of the Soviet Union in 1989 and the euphoria around the globalisation of American-style consumerism, *Limits to Growth* was more or less forgotten. Its warnings were cast aside in favour of the consensus celebrated by Francis Fukuyama in his 1992 book *The End of History*: free-market capitalism was the only game in town, and it seemed for all the world that it was going to last for ever.

*

But then something changed. With the global financial crisis of 2008 the party came crashing to an end. People's faith in the limitless magic of the free market and the universal promise of the American Dream was shaken to its core. Major banks collapsed, and millions of people around the world lost their homes and jobs. In a desperate bid to get growth going again, many governments bailed out the banks, gave tax breaks to the rich, slashed labour laws, and cut social spending with harsh austerity measures. This triggered waves of popular social movements: Occupy Wall Street, the Indignados, the Arab Spring – people angry at a system that prioritises capital over people. And all of this was unfolding as the world began to wake up to the reality of climate change, with storms, fires, droughts and flooding capturing headlines on a regular basis.

Against the backdrop of systemic crises, people have begun to interrogate the prevailing economic consensus, and the question of ecological limits has come rushing to the fore again. This time, however, the old *Limits to Growth* mindset has been supplanted by a completely new way of thinking about limits.

The problem with the *Limits to Growth* report is that it focused only on the finite nature of the resources that we need to keep the economy running. This way of thinking about limits is vulnerable to those who point out that if we can find new reserves, or substitute new resources for old, and if we develop methods of improving the yields of renewable resources, then we don't have to worry about those limits. Sure, this process of substitution and intensification can only go so far – at some point we'll reach an absolute limit – but for all we know that could be a long way off.

But this isn't how ecology actually works. The problem with economic growth isn't just that we might run out of resources at some point. The problem is that it progressively degrades the integrity of ecosystems. As onshore oil reserves run dry we can switch to offshore reserves, but both sources contribute to climate breakdown. We might be able to substitute one metal for another, but ramping up the mining of any metal is going to poison rivers and ruin habitats. And we might be able to intensify our extraction from the land by pumping it full of chemicals, but not without triggering soil depletion and pollinator collapse. The process of substitution and intensification might get us around resource limits for a while, but it still drives ecological breakdown. That's the problem.

In recent years, ecologists have developed a new, more scientifically robust way of thinking about limits. In 2009, a team led by Johan Rockström at the Stockholm Resilience Centre, the US climatologist James Hansen, and Paul Crutzen, the man who coined the term Anthropocene, published a groundbreaking paper describing a new concept they referred to as 'planetary boundaries'.[31] The Earth's biosphere is an integrated system that can withstand significant pressures, but past a certain point it begins to break down. Drawing on data from Earth-systems science, they identified nine potentially destabilising processes that we have to keep under control if the system is to remain intact: climate change, biodiversity loss, ocean acidification, land-use change, nitrogen and phosphorous loading, freshwater use, atmospheric aerosol loading, chemical pollution and ozone depletion.

Scientists have estimated 'boundaries' for each of these processes. For example, atmospheric carbon concentration should not breach 350ppm if the climate is to remain stable (we crossed that boundary in 1990, and hit 415ppm in 2020); the extinction rate should not exceed ten species per million per year; conversion of forested land should not exceed 25% of the Earth's land surface; and so on. These boundaries aren't 'hard' limits, in the strict sense. Crossing them doesn't mean that the Earth's systems will immediately shut down. But it does mean we are entering a danger zone where we risk triggering tipping points that could eventually lead to irreversible collapse.

In terms of ecology, this is a more coherent way of thinking about limits. Our Earth is a plentiful place – it generates an abundance of forests and fish and crops every year. It is also remarkably resilient, as it not only reproduces these things as we use them, it absorbs and processes our waste too: our emissions, our chemical run-off, and so on. But in order for the planet to maintain these capacities, we can only take as much as its ecosystems can regenerate, and pollute no more than the atmosphere and rivers and soil can safely absorb. If we overshoot these boundaries, ecosystems begin to break down and the web of life begins to unravel. That's what's happening right now. According to the most recent data, we have already shot past four of the planetary boundaries: for climate change, biodiversity loss, deforestation and biogeochemical flows. And ocean acidification is nearing the boundary.

So what does all this mean for economic growth? Hitting or crossing the planetary boundaries doesn't mean that economic growth will suddenly stop. We are already sliding into dangerous tipping points, and growth shows no sign of ending. In fact, one can imagine that GDP might continue growing even as social and ecological systems begin to collapse. Capital will pile into new growth sectors like sea walls, border militarisation, Arctic mining and desalinisation plants. Indeed, many of the world's most powerful governments and corporations are already positioning themselves to capitalise on likely disaster scenarios. They know very well what's ahead if we carry on with business as usual.

Of course, as a strategy for maintaining aggregate GDP growth, this will only work for a time. As ecological breakdown triggers tipping points, as agricultural output declines, as mass displacement undermines political stability, and as cities are ruined by rising seas, the environmental, social and material infrastructure that underpins the possibility of growth – and indeed the possibility of organised civilisation – will fall apart.

Trying to predict when we might bump into the limits to growth is exactly the wrong way to think about it. We will find ourselves plunging into ecological collapse well before we run into the limits to growth. Once we realise this, it completely changes the way we think about the question of limits. As the political ecologist Giorgos Kallis has put it, the problem isn't that there are near-term limits to growth – it's that there *aren't*. If we want to have any chance of surviving the Anthropocene, we can't just sit around and wait for growth to crash into some kind of external limit. We must choose to limit growth ourselves. We need to reorganise the economy so that it operates within planetary boundaries, to maintain the Earth's life-supporting systems which we depend on for our existence.[32]

Three

Will Technology Save Us?

> Climate change is an engineering problem and has engineering solutions.
>
> Rex Tillerson, former CEO of ExxonMobil

Even as the evidence about the relationship between economic growth and ecological breakdown continues to pile up, growthism remains entrenched. It has the staying power and ideological fervour of a religion. Of course, this is hardly surprising: our economic system is structurally dependent on growth, it serves the interests of the most powerful factions of our society, and it is rooted in a deep-seated world view of dominion and dualism that goes back some 500 years. This edifice will not yield easily. Not even to science.

When I reflect on the conflict between science and growthism, I can't help but think of Charles Darwin. As I mentioned in the introduction, Darwin's findings about evolution posed such a radical challenge to the dominant world view at his time that they were almost impossible for people to accept. To see humans as descended from non-humans rather than created in the image of God required a total paradigm shift. Something similar is happening right now. Ecological science requires that we learn to see the human economy not as separate from ecology but as embedded within it. This poses a radical challenge to the dominant world view, and to capitalism itself. Yet rather than accept this evidence and change their world view, those who seek to preserve the present system instead devise elaborate alternative theories explaining that we needn't change course; that we can carry on growing the global economy indefinitely and everything will be fine.

This narrative relies heavily on the claim that technology will save us, in one way or another. For some, it is a simple matter of switching the global economy to renewable energy and electric cars; once we do that, there's no reason we can't keep growing for ever. After all, solar and wind power are getting cheaper all the time, and Elon Musk has shown that it's possible to mass-produce storage batteries at a rapid clip. For others, it's a matter of 'negative-emissions technologies' that will pull carbon out of the atmosphere. Still others bank on the hope of enormous geo-engineering schemes: everything from blocking out the sun to changing the chemistry of the oceans. Of course, even if these solutions succeed in stopping climate change, continued growth will still drive continued material use, and continued ecological breakdown. But here too some insist that this is not a problem. Efficiency improvements and recycling technologies will allow us to make growth 'green'.

These hopes have been touted by some of the richest and most powerful people in the world, including presidents and billionaires. The ecological crisis is no reason to start questioning the economic system, they say. It's a comforting narrative, and one I myself once clung to. But the more I have explored these claims, the more it has become clear to me that to take this position requires accepting an extraordinary risk. We can choose to keep shooting up the curve of exponential growth, bringing us ever closer to irreversible tipping points in ecological collapse, and hope that technology will save us. But if for some reason it doesn't work, then we're in trouble. It's like jumping off a cliff while hoping that someone at the bottom will figure out how to build some kind of device to catch you before you crash into the rocks below, without having any idea as to whether they'll actually be able to pull it off. It might work . . . but if not, it's game over. Once you jump, you can't change your mind.

If we're going to take this approach, the evidence for it had better be rock-solid. We'd better be dead certain it will work.

Gambling in Paris

Everyone heaved a collective sigh of relief on the night that the world's governments finally came to an agreement on climate change. It was Paris in 2015, and despite the cold darkness of December the city felt bright and hopeful. The Eiffel Tower bore the words '1.5 degrees' in giant glowing letters. It was a heartening moment – a welcome sign that our leaders were finally willing to take the difficult steps necessary to avert climate catastrophe, after many decades of failure. And in the years since that thrilling December night it's been easy to assume that we must be more or less on track.

Here's how the Paris Agreement works. Each country submits a pledge on how much they will reduce their annual emissions. The pledges – known as Nationally Determined Contributions – are supposed to be set in line with the goal of keeping warming to 1.5°C. But if you add up all the pledges that have been made by signatory nations as of 2020, you'll notice something rather strange: they don't come anywhere close to keeping us under 1.5°C. In fact, they don't even keep us under 2°C. Even if all the countries in the world fulfil their pledges – which are voluntary and non-binding, so there's certainly no guarantee of this – global emissions will *keep rising.* We'll still be hurtling towards 3.3°C of global warming by the end of the century. In other words, even with the Paris Agreement in place, we're on track for catastrophe.

What's going on here? How is it possible that emissions will keep rising even under a plan that's meant to cut them? And why does nobody seem to be worried about this?

There's a backstory. In the early 2000s, IPCC modellers realised that the emissions reductions required to keep climate change under control were so steep that they were likely to be incompatible with continued economic growth. Growing the global economy means growing energy demand, and growing energy demand makes the task of transitioning to clean energy significantly more difficult. As long as energy demand keeps going up it's unlikely we'll be able to roll out enough clean energy to cover it in the short time we have left. As far as anyone could tell, the only feasible way to do it would be to actively slow down industrial production. Reducing the scale of global energy use would make it easier to accomplish a quick transition to renewables.

But policymakers knew this conclusion wouldn't go down well, and they feared it would be a tough sell in international negotiations. The idea of a trade-off between economic growth and climate action would make it impossible to get key nations like the United States on board, and could ultimately scupper any chances of securing an international agreement on climate change. The risks were just too high. Countries were also rallying around the goal of ending global poverty, and world leaders kept saying that the only effective way to end poverty is to ramp up global economic growth. The idea that climate mitigation might come with trade-offs for growth would be impossible to swallow. Growth is like the third rail: touch it and you die. Growth must go on.

Fortunately, they found a solution. Or so it seemed.

*

In 2001, an Austrian academic named Michael Obersteiner published a paper describing a brilliant new technology: an energy system that would not only be carbon-neutral, but would actively pull carbon out of the atmosphere.[1] The proposal was stunning in its elegance. First you establish massive tree plantations around the world. The trees suck CO_2 out of the atmosphere as they grow. Then you harvest the trees, churn them into pellets, burn them in power plants to generate energy, capture the carbon emissions at the chimneys and store it all underground where it can never escape. Voila: a global energy system that produces 'negative emissions'.

This technology is known as BECCS: bio-energy with carbon capture and storage. When Obersteiner published his paper there was no evidence that the scheme would actually work; it was just speculation. But the sheer possibility of it captivated those who were looking for politically palatable ways of staying under 2°C. The idea was that we can get by with making relatively minor reductions to CO_2 emissions – nothing that would pose any significant threat to economic growth – so long as we manage to get BECCS up and running. We'll overshoot the carbon budget, but that's OK because BECCS will pull the excess carbon back out of the atmosphere later in the century, bringing us back into the safety zone. Emit now, clean up later.

It was a crazy gamble, and everyone knew it. But the idea spread like wildfire. It held out the tantalising possibility of meeting our climate goals while keeping capitalism intact, and while allowing rich nations, who wield so much power in the climate negotiations, to maintain their high levels of consumption. It was incredibly alluring – a kind of get-out-of-jail-free card – and it offered real hope to green growth optimists.

A few years after Obersteiner's paper was published the IPCC started including BECCS in its official models, even though there was still no evidence of its feasibility. And in 2014 the idea took centre stage: BECCS appeared in the IPCC's Fifth Assessment Report (AR5), not only as a side show, but as the dominant assumption in no fewer than 101 of the 116 scenarios for staying under 2°C. AR5 is the blueprint that the Paris Agreement relies on. Governments are using the AR5 scenarios as a guide when it comes to deciding how quickly to reduce their emissions. This helps explain why national plans significantly overshoot the carbon budget for 2°C: it's because everyone's relying on scenarios that assume BECCS will save us.

In other words, BECCS sits right at the centre of our big plan to save the world, even though most people have never even heard of it. Journalists never mention it, our politicians never talk about it; not because they're trying to hide something, or because it's too complicated to explain, but because most of them don't know it even exists. They're just following the scenarios. The future of our planet's biosphere, and of human civilisation, hinges on a plan that very few people know about, and to which nobody has consented.

Jumping off a cliff

But there's a hitch. Climate scientists have been sounding the alarm about BECCS from day one, and their objections have grown louder with every passing year. There are four main problems with the idea – each potentially fatal.

First, BECCS has never been proven to be scalable. To make it work would require that we create a global carbon-capture-and-storage (CCS) system capable of sucking up some 15 billion tons of CO_2 a year. Right now we have capacity to handle about 0.028 billion tons – and only a fraction of that is verified. Since a typical CCS facility can handle about 1 million tons, we would need to construct some 15,000 new facilities all around the world.[2] The scale of this development is enormous – it would be one of the biggest infrastructural feats ever attempted in human history – and we have no idea whether it's possible to pull it off in time. We also have no idea whether it will be commercially viable. Right now it is not. It will only become viable if governments around the world agree to put a price on carbon at least ten times higher than it is presently priced in the European Union.[3]

This isn't an insurmountable obstacle, but it does make the 'overshoot now, clean up later' strategy highly risky. If we bet on BECCS, and choose to not reduce our emissions in the near term, there's no going back. If BECCS fails then we will be locked into a future of extreme global warming. When you're gambling with the fate of human civilisation, and indeed with the web of life itself, the stakes are simply too high.

In 2014, the year before the Paris climate summit, fifteen scientists penned a letter warning against BECCS in the pages of the prestigious academic journal *Nature Climate Change*. They argued that the widespread use of BECCS in the climate models 'might become a dangerous distraction' from the imperative of reducing emissions.[4] And they're not alone. The following year, another forty scientists argued that reliance on negative-emissions technologies like BECCS is 'extremely risky'.[5] Professor Kevin Anderson of Manchester University, one of the world's leading climate scientists, has been a particularly vocal critic of BECCS. In a 2016 article in the journal *Science,* he argued that the Paris Agreement's reliance on BECCS is 'an unjust and high-stakes gamble'.[6] Dozens of other scientists are coming forward with the same conclusion.

Even if we somehow manage to overcome the technical and economic obstacles, we'll bump straight into another crisis. In order for BECCS to remove as much carbon as the IPCC scenarios assume, we will need to create biofuel plantations covering an area two to three times the size of India, gobbling up about two-thirds of the planet's arable land. This would require shifting land away from food crops, which is a problem when we're trying to feed a population that's on track to grow to at least 9 billion by the middle of the century. In other words, relying on BECCS at scale would be likely to cause severe food shortages and could even trigger famines. It's not difficult to imagine the conflicts this would catalyse. And let's not pretend that powerful nations are going to willingly give their own land over to biofuels; it's more likely they'll attempt to seize land elsewhere, setting off a kind of climate colonialism. Where wars were once fought over access to oil, they would instead be fought over land for biofuels.

On top of all this, BECCS would be an ecological disaster in its own right. A team of researchers led by the German scientist Vera Heck has estimated that the rollout of biofuel plantations at scale would have a number of devastating impacts. Vast tracts of forest would have to be destroyed, slashing global forest cover by

10% from its already-precarious levels. This would drive an add-itiónal 7% loss in biodiversity, further exacerbating mass extinction.[7] And the use of chemical fertilisers for monoculture on such an unprecedented scale would decimate insect popula-tions, pollute water systems, exacerbate soil depletion and worsen coastal dead zones.[8] In addition, BECCS plantations would require twice as much water as we already use for farm-ing, placing communities and ecosystems around the world under significant stress.[9]

In other words, BECCS might help us in the battle against cli-mate change, but only by pushing us headlong into a number of other deadly problems. If global warming was the only crisis we were facing, this might seem like a reasonable risk to take. But given that it's only one part of a broader ecological crisis, it doesn't make any sense. It's a suicidal strategy.

In addition – and here's the final nail in the coffin – even if by some miracle we managed to avoid all of these complications and get BECCS working smoothly we would still be in trouble, because overshooting the carbon budget means triggering possible tipping points and feedback loops that could push tem-peratures completely out of our control. And if that happens, the whole exercise would have been in vain. We might be able to pull carbon out of the atmosphere at some future point, but we can-not reverse climate tipping points.[10]

*

It is worrying that much of the world has been devising climate strategy around such a dangerous and uncertain technology. In fact, Obersteiner himself – the original inventor of the BECCS concept – has expressed concern about the use of his idea. He says he conceived of BECCS purely as a 'risk-management strategy', or a 'backstop technology' in case climate feedback loops turn out to be worse than we expect. He saw it as something we could use to help us reach our emissions targets under emergency conditions. Modellers have 'misused' the idea, he says, by including it in regular scenarios for staying under 1.5 or 2°C. Afraid of calling for steeper emissions cuts, policymakers have been using BECCS as an excuse to carry on with the status quo. Some of the other key figures behind early articulations of BECCS have also raised questions, pointing out that the technology was only ever meant to be used on a small scale. They warned from the beginning that a large-scale rollout would be a social and ecological disaster – and yet modellers have run with it anyhow.[11]

The scientific consensus against BECCS is now rock-solid. In early 2018, the European Academies' Science Advisory Council, a body that brings together the national science academies of all the states of the European Union, published a report condemning the reliance on BECCS and other negative emissions technologies. In the scientific community, it's difficult to get a stronger conclusion than this. The report urges that we stop speculating on tech fantasies and get serious about deep and aggressive cuts to emissions.

This isn't to say that BECCS will have no part to play in our battle against climate breakdown. It will have to be part of the mix, and we should invest in research and testing. But we need to face up to the fact that it can't be rolled out on anything near the scale that modellers propose. The latest assessments show that safe use of BECCS – in a way that respects planetary boundaries and human food systems – will allow us to reduce global emissions by at most 1%. That's an important contribution, to be sure; but it's a far cry from the saviour technology that people once hoped it would be.[12]

The fight for 1.5

The IPCC has been paying attention to these critiques. In October 2018, it released a special report outlining what it will take for us to keep global warming under 1.5°C if we accept that we cannot reasonably rely on negative emissions technologies. The report landed like a bombshell in the world's media. It was difficult to find an outlet that didn't carry the headline findings: if we want to have a decent shot at keeping temperatures under 1.5°C, we have to cut global emissions in half by 2030 and get to zero before 2050.

It is impossible to overstate how dramatic this trajectory is. It means nothing less than the rapid and dramatic reversal of our present direction as a civilisation. We have built up a global fossil-fuel infrastructure over the past 250 years, and now we have to completely overhaul it in only thirty. Everything has to change, in a matter of decades. And keep in mind that this is for the world as a whole. Rich nations have to cut emissions much more quickly, given the scale of their historical contributions to climate breakdown, while poorer nations can take it more slowly. Scientists at the Stockholm Environment Institute calculate that rich countries need to reach zero emissions before 2030.[13]

The IPCC report had a galvanising effect, spurring citizens to action. Students staged climate strikes across Europe and North America. In London, the Extinction Rebellion movement blockaded five bridges across the River Thames, demanding that the UK government act immediately to achieve rapid emissions reductions. Opinion polls showed that a large majority of the British public supported the movement's aims. Over the following months, the political conversation changed in ways that nobody would have expected. Parliament declared a climate emergency, and accepted a legally binding target for reducing emissions to zero by 2050. While this target fails to meet the earlier decarbonisation dates required of rich nations, it nonetheless marked a significant shift.

Meanwhile, a similar movement was rippling across the United States. In February 2019, Congresswoman Alexandria Ocasio-Cortez and Senator Edward Markey released a resolution for the Green New Deal, which called for a ten-year national mobilisation with the goal of shifting the United States to 100% clean energy. The idea caught fire: the progressive wing of the Democratic Party lined up behind it, and opinion polls showed that more Americans supported the idea than rejected it. Republican leaders rounded on the plan, and conservative media launched relentless attacks. But for the first time the nation was having an open conversation about serious climate policy – something that seemed unthinkable for a country where climate denialism has so long been entrenched.

Green growth?

All this brings us into new political terrain. A new consensus has emerged. While for decades we have been relying on market mechanisms to somehow magically fix the climate crisis, it's now clear this approach isn't going to do. The only way to make it work is with co-ordinated government action on a massive scale. Proponents of the Green New Deal have it right: we need to pump public investment into building renewable energy infrastructure at a historically unprecedented rate, reminiscent of the industrial retooling that enabled the Allies to win the Second World War.

But there's something troubling about the way this idea has been picked up and repackaged by some media pundits. The claim is that transitioning to clean energy will liberate capitalism from any concerns about ecology. It will pave the way to 'green growth', they say, and we can keep expanding the economy for ever. It's a compelling story. It seems so obvious and straightforward. And not surprisingly, it has seized the imaginations of orthodox economists and politicians. But this narrative suffers from a number of serious flaws. In fact, scientists go so far as to reject green growth hopes as empirically baseless.

The key point to grasp is that while it's possible to transition to 100% renewable energy, we cannot do it fast enough to stay under 1.5°C or 2°C if we continue to grow the global economy at existing rates. Again: more growth means more energy demand, and more energy demand makes it all the more difficult (and probably impossible) to generate enough renewable capacity to meet it in the short time we have left.

Don't get me wrong. We have made extraordinary gains in renewable energy capacity over the past couple of decades, and

139

this is wonderful news. Today the world is producing 8 billion more megawatt hours of clean energy each year than in 2000. That's a lot – enough to power all of Russia. But over exactly the same period, economic growth has caused energy demand to increase by 48 billion megawatt hours. In other words, all the clean energy we've been rolling out covers only a fraction of new demand. It's like shovelling sand into a pit that just keeps getting bigger. Even if we doubled or tripled the output of clean energy production, we would still make zero dent in global emissions. Growth keeps outstripping our best efforts to decarbonise.

Think about it this way. If we continue to grow the global economy at projected rates, it will more than double in size by the middle of the century – that's twice as much extraction and production and consumption than we are presently doing, all of which will suck up nearly twice as much end-use energy than would otherwise be the case.[14] It will be unimaginably difficult for us to decarbonise the *existing* global economy in the short time we have left; impossible to do it nearly twice over. It would require that we decarbonise at a rate of 7% per year to stay under 2°C (which is dangerous), or 14% per year to stay under 1.5°C. That's two to three times faster than what scientists say is possible even under best-case scenario conditions.[15] As one team of researchers put it, it is 'well outside what is currently deemed achievable'.[16]

Our insistence on perpetual growth is making our task much more difficult than it needs to be. It's as though we've chosen to fight this life-or-death battle facing uphill, blindfolded, with our hands tied behind our backs. We are knowingly stacking the odds against ourselves.

This conclusion is shared widely among scientists, including at the very highest levels. Even the IPCC itself acknowledges that without BECCS and other speculative technologies, there's no feasible way to roll out clean energy fast enough to get to zero emissions by 2050 as long as energy demand keeps growing.[17] If we want to succeed, we have to do exactly the opposite: we have to scale down energy use.

*

Even if this wasn't a problem, there's yet another issue we have to face up to – to do with clean energy itself. When we hear the phrase 'clean energy' it normally calls to mind happy, innocent images of warm sunshine and fresh wind. But while sunshine and wind are obviously clean, the infrastructure we need to capture it is not. Far from it. The transition to renewables is going to require a dramatic increase in the extraction of metals and rare-earth minerals, with real ecological and social costs.

In 2017, the World Bank released a report offering the first comprehensive look at this question.[18] Researchers modelled the increase in material extraction that would be required to build enough solar and wind utilities to produce an annual output of about 7 terawatts of electricity by 2050. That's enough to power a bit less than half of the global economy. By doubling the World Bank figures, we can estimate what it will take to get all the way to zero emissions (not including a little bit of hydropower, geothermal and nuclear to top it off) – and the results are staggering: 34 million metric tons of copper, 40 million tons of lead, 50 million tons of zinc, 162 million tons of aluminium, and no less than 4.8 billion tons of iron.

In some cases, the transition to renewables will require a massive increase over existing levels of material extraction. For neodymium – an essential element in wind turbines – extraction will need to rise by nearly 35% over current levels. Higher-end estimates reported by the World Bank suggest it could double. The same is true of silver, which is a critical ingredient in solar panels. Silver extraction will go up 38% and perhaps as much as 105%. Demand for indium, also essential to solar technology, will more than triple and could end up skyrocketing by 920%.

And then there are all the batteries we're going to need for power storage. To keep energy flowing when the sun isn't shining and the wind isn't blowing will require enormous batteries at the grid level. This means 40 million tons of lithium – an eye-watering 2,700% increase over current levels of extraction.

That's just for electricity. We also need to think about vehicles. In 2019, a group of leading British scientists submitted a letter to the UK's Committee on Climate Change outlining their concerns about the ecological impact of electric cars.[19] They agree, of course, that we need to end the sale and use of combustion engines and switch to electric vehicles as quickly as possible. But they pointed out that replacing the world's projected fleet of 2 billion vehicles is going to require an explosive increase in mining: global annual extraction of neodymium and dysprosium will go up by another 70%, annual extraction of copper will more than double, and cobalt will need to increase by a factor of almost four – all for the entire period between now and 2050. We need to switch to electric cars, yes; but ultimately we need to radically reduce the number of cars we use.

The problem here is not that we're going to run out of key minerals – although that may indeed become a concern. The real issue is that this will exacerbate an already existing crisis of overextraction. Mining has already become a big driver of deforestation, ecosystem collapse and biodiversity loss around the world. If we're not careful, growing demand for renewable energy will exacerbate this crisis significantly.

Take silver, for instance. Mexico is home to the Peñasquito mine, one of the biggest silver mines in the world. Covering nearly 40 square miles, the operation is staggering in its scale: a sprawling open-cast complex ripped into the mountains, flanked by two waste dumps each a mile long, and a tailings dam full of toxic sludge held back by a wall that's 7 miles around and as high as a fifty-storey skyscraper. This mine will produce 11,000 tons of silver in ten years before its reserves, the biggest in the world, are gone.[20] To transition the global economy to renewables, we need to commission up to 130 more mines on the scale of Peñasquito. Just for silver.

Lithium is another ecological disaster. It takes 500,000 gallons of water to produce a single ton of lithium. Even at present levels of extraction this is causing real problems. In the Andes, where most of the world's lithium is located, mining companies are burning through the water tables and leaving farmers with nothing to irrigate their crops. Many have had no choice but to abandon their land altogether. Meanwhile, chemical leaks from lithium mines have poisoned rivers from Chile to Argentina, Nevada to Tibet, killing off whole freshwater ecosystems. The lithium boom has barely started, and it's already a catastrophe.[21]

And all of this is just to power the global economy by 2050. Things become even more extreme when we start accounting for growth into the future. As energy demand continues to rise, material extraction for renewables will become all the more aggressive – and the more we grow, the worse it will get. Even after achieving a full energy transition, to keep the global economy growing at projected rates would mean doubling the total global stock of solar panels, wind turbines and batteries every thirty or forty years, for ever.

It's important to keep in mind that most of the key materials for the energy transition are located in the global South. Parts of Latin America, Africa and Asia are likely to become the target of a new scramble for resources, and some countries may become victims of new forms of colonisation. It happened in the sixteenth, seventeenth and eighteenth centuries with the hunt for gold and silver from South America. In the nineteenth century, it was land for cotton and sugar plantations in the Caribbean. In the twentieth century, it was diamonds from South Africa, cobalt from the Democratic Republic of Congo, and oil from the Middle East. It's not difficult to imagine that the scramble for renewables might become similarly violent.

If we don't take precautions, clean energy firms could become as destructive as fossil fuel companies – buying off politicians, trashing ecosystems, lobbying against environmental regulations, even assassinating community leaders who stand in their way, a tragedy that is already unfolding.[22] This is important. Progressives who promote the idea of a Green New Deal or other plans for rapid energy transition also tend to promote values of social and ecological justice. If we want the transition to be just, we need to recognise that we cannot increase our use of renewable energy indefinitely.

Some hope that nuclear power will help us get around these problems – and surely it will need to be part of the mix. But nuclear comes with its own constraints. The main problem is that it takes so long to get new power plants up and running that they can play only a small role in getting us to zero emissions by the middle of the century. Even in the longer term, some scientists worry that nuclear can't be scaled up beyond about 1 terawatt.[23] Moreover, if for whatever reason we don't manage to stabilise the climate – a real possibility – nuclear sites will be vulnerable to severe storms, rising seas and other disasters that could turn them into radiation bombs. With climate breakdown bearing down on us, relying too much on nuclear could become a dangerous gamble.

As for fusion power – the running joke is that engineers have been saying it's a decade away for about six decades now. While we have managed to create successful fusion reactions, the problem is that the process requires more energy than it produces. A big fusion experiment presently under way in France may be close to solving that problem (and that's a big maybe), but even the most optimistic projections indicate that it won't happen for another ten years. It would take another decade after that to get fusion power to the grid, and many more decades to scale it up. So while the prospects are exciting, the record so far is not encouraging, and in any case the timeline is too long. We may have fusion power sometime this century, but we certainly can't rely on it to keep us within the safe carbon budget. Without a miraculous technological breakthrough, the energy transition is going to need to focus mostly on solar and wind.

None of this is to say we shouldn't pursue a rapid transition to renewable energy. We absolutely must, and urgently. But if we want the transition to be technically feasible, ecologically coherent and socially just, we need to disabuse ourselves of the fantasy that we can carry on growing aggregate energy demand at existing rates. We must take a different approach.

The planet remade

In the face of this evidence, those who insist on continued growth have been turning to increasingly outlandish ideas – not just BECCS but a growing menu of science-fiction techno-fixes based on large-scale geo-engineering. Most of these schemes are so difficult and expensive to implement that you might as well just swallow the cost of actually reducing emissions instead. But there's one that stands out from the crowd, and which has attracted significant attention. It's called solar radiation management.

The idea is to use a fleet of jets to inject aerosols into the stratosphere, forming a giant veil around the Earth to reflect sunlight and therefore cool the planet. It's relatively cheap and easy to do. So easy, in fact, that scientists worry that rogue agents – say, a meddling billionaire or a desperate island state that's about to go underwater – could pull it off single-handedly. A number of governments are commissioning research on solar radiation management, and the idea has been celebrated by fossil fuel executives who see it as a way to preserve their business model.

But it's not without its risks. Existing models suggest it could end up tearing holes in the ozone layer, slow photosynthesis to the point of decreasing crop yields, and irreversibly alter global rainfall patterns and weather systems – mostly to the detriment of the global South. Jonathan Proctor, a scientist who studies solar radiation management, says 'the side effects of treatment are as bad as the original disease'. Janos Pasztor, another expert in this field, points out that the consequences could end up being even worse than we're able to predict: 'The global atmosphere is unbelievably complex ... we have advanced computer modelling with supercomputers, but we still don't really know how to model it.'[24]

Perhaps the biggest problem, though, is that aerosols don't last long in the stratosphere, so for the plan to work that fleet of jets would have to be at it constantly. And if for whatever reason they stopped, we'd be in real trouble: global temperatures would shoot up again at a rapid pace, rising several degrees within a single decade. This sudden heating, known as 'termination shock', would leave countries with little time to adapt. Ecosystems would fall under tremendous strain and huge numbers of species would be wiped out.[25] Scientists regard this approach to be too risky to implement, and – like all geo-engineering schemes – a dangerous distraction from the objective of cutting emissions fast.

It's worth pausing to reflect on the growing fascination with geo-engineering. What's interesting about it is that it embodies the very same logic that got us into trouble in the first place: the idea that the living planet, rendered as mere 'nature', is nothing but a set of passive materials that can be subdued, conquered and controlled. Geo-engineering represents dualism taken to astonishing new extremes, unimaginable by Bacon and Descartes, where the planet itself must be bent to the will of man so that capitalist growth can continue indefinitely. The fatal flaw of geo-engineering is that it seeks to solve the ecological crisis with the very same thinking – the very same hubris – that created it in the first place. But perhaps more immediately, the problem with geo-engineering is that it is ecologically incoherent. Solar radiation management is only a partial response to the crisis we face. It would do nothing to slow the pace of ocean acidification, or deforestation, or soil depletion, or mass extinction. And this brings us to the next point.

Out of the frying pan, into the flames

Let's pretend, just for the sake of argument, that none of this was a problem. Put aside the evidence for a moment and imagine that we somehow manage to achieve a rapid transition to clean energy while still growing the global economy, and that we can continue growing energy demand indefinitely without worrying about the material extraction it will entail or the pressure it will place on already-exploited regions of the world. Let's say we invent fusion power tomorrow and scale it up in a decade. Surely such a scenario meets the requirements for green growth, right?

The problem with this vision is that it misses one key, unavoidable point: emissions are only one part of the crisis. In addition to climate breakdown, we are already overshooting a number of other planetary boundaries, driven by ever-increasing extraction from the Earth. The problem isn't just the type of energy we're using; it's what we are doing with it.

Even if we had a 100%-clean-energy system, what would we do with it? Exactly what we are doing with fossil fuels: raze more forests, trawl more fish, mine more mountains, build more roads, expand industrial farming, and send more waste to landfill – all of which have ecological consequences our planet can no longer sustain. We will do these things because our economic system demands that we grow production and consumption at an exponential rate. In fact, the whole idea behind using clean energy to power a 'green growth' system is so that we can keep growing material production and consumption. Otherwise why would we need to keep growing energy demand?

Switching to clean energy will do nothing to slow down all these other forms of ecological breakdown. Escaping the frying pan of climate disaster doesn't help us much if we end up hopping into the flames of ecological collapse.

*

Proponents of green growth have a quick response, however. They insist that all we need to do is 'decouple' GDP growth from resource use. There's no reason we can't just dematerialise economic activity, and keep growing GDP even as resource use falls back down to sustainable levels. They admit, of course, that resource use has historically gone up in lockstep with GDP. But that's at a *global* level. If we look at what's happening in certain high-income nations, which are becoming more technologically sophisticated and rapidly shifting from manufacturing to services, we might find clues to what the future could hold.

When this idea was first floated, it appeared that there was indeed some interesting evidence to back it up. Green growth proponents pointed out that the 'domestic material consumption' (DMC) of Britain, Japan and a number of other rich countries has been decreasing since at least 1990, even as GDP has continued to grow. Even in the United States, DMC has more or less flattened out over the past couple of decades. This data was picked up by journalists who were quick to announce that rich countries had reached 'peak stuff' and were now 'dematerialising' – proof that we can keep growing GDP for ever without having to worry about ecological impact.

But ecologists have long rejected these claims. The problem with DMC is that it ignores a crucial piece of the puzzle: while it includes the imported goods a country consumes, it does not include the resources involved in producing those goods. Because rich countries have outsourced so much of their production to other countries – mostly in the global South – that side of resource use has been conveniently shifted off their balance sheet. To account for this, scientists prefer to use a measure called 'material footprint', which includes the total resources embodied in a nation's imports.

Using this more holistic measure it quickly becomes clear that the material consumption of rich nations hasn't been falling at all. In fact, in recent decades it's been increasing dramatically, even to the point of outpacing GDP growth. There has been no decoupling. It was all an illusion of accounting.[26]

As it turns out, the much-celebrated shift to services has delivered no improvements at all when it comes to the resource intensity of rich nations. Services represent 74% of GDP in high-income nations, having grown rapidly since deindustrialisation began in the 1990s, and yet the material use of high-income nations is outpacing GDP growth. Indeed, while high-income nations have the highest share of services in terms of contribution to GDP, they also have the highest per capita material footprints. By far. The same is true on a global scale. Services have grown from 63% of GDP in 1997 to 69% in 2015, according to World Bank data. Yet during this same period global material use has accelerated. In other words, we have seen a *rematerialisation* of the global economy even as we have shifted to services.

What explains this strange result? It's partly that the incomes people make in the service economy end up getting used to buy material goods. People might earn their money on YouTube, but they spend it buying things like furniture and cars. But it's also because services themselves turn out to be resource-intensive in their own right. Take the tourism sector, for example. Tourism is classed as a service, and yet it requires an enormous material infrastructure to keep it going – airports, planes, buses, cruise ships, resorts, hotels, swimming pools and theme parks (all of which are services themselves).

Given the data we have so far, there's no reason to believe that shifting to services is somehow magically going to reduce our resource use. It's time to put that myth aside.

There's also something else going on. With every year that goes by, it becomes more and more difficult to extract the same amount of materials from the earth. All the stuff that's close to the surface and easy to get to has already been snatched up. As we exhaust easily accessible reserves of minerals and metals we have to dig ever deeper and more violently to get more. We know that oil companies are being forced to turn to fracking, deep-sea drilling and other 'tight plays' to reach remaining oil reserves, using up more energy and materials to get the same amount of fuel. The same thing is happening with mining. According to the UN Environment Programme (UNEP), today three times more material has to be extracted per unit of metal than a century ago.[27] Part of this is also down to the decline in the quality of metal ore – by as much as 25% over the past ten years alone – meaning we need to extract and process more ore just to get the same quantity of finished product.[28] In other words, despite significant improvements in mining technology, the material intensity of mining has been getting *worse*, not better. And UN scientists say this troubling trend will only continue.

When faced with this data, proponents of green growth double down. That's all in the past, they say. Just because it hasn't been done before doesn't mean it's not possible. We can still change our future direction. We just need to roll out the right technology and the right policies. Governments can impose taxes on resource extraction while at the same time investing in efficiency improvements. Surely this will shift patterns of consumption towards goods that are less resource intensive? People will spend their money on movies and plays, for example, or on yoga and restaurants and new computer software. So GDP will continue growing for ever while resource use declines.

It's a comforting thought, and it sounds reasonable enough. Fortunately, we now have the evidence to test whether it holds up. Over the past few years scientists have developed a number of models to determine the impact of policy changes and technological innovation on material use. And the results are quite surprising.

*

The first study was published in 2012 by a team of scientists led by the German researcher Monika Dittrich.[29] The group ran a sophisticated computer model showing what would happen to global resource use if economic growth continued its current trajectory, at about 2 to 3% a year. The scientists found that human consumption of materials would rise at exactly the same rate as GDP. Using current data, that means hitting over 200 billion tons by 2050 – four times over the safe boundary. Disaster.

Then the team re-ran the model to see what would happen if every nation in the world immediately adopted best practice in efficient resource use – an extremely optimistic assumption. The results improved: resource consumption rose more slowly. But it still rose. When resource use rises more slowly than GDP, that's called *relative* decoupling. But it's a far cry from the sufficient *absolute* decoupling we need. So, no green growth.

In 2016, a second team of scientists tested a different scenario: one in which the world's nations all agreed to go above and beyond existing best practice.[30] In their best-case scenario, they assumed a tax that would raise the price of carbon to $236 per ton (which in turn raises the costs of material extraction and transportation), and imagined technological innovations that would double the efficiency with which we use resources. The results were almost exactly the same as in Dittrich's study. Even under these stringent conditions, resource use keeps going up. No absolute decoupling, and no green growth.

Finally, in late 2017 the UNEP – an institution that once eagerly promoted green growth theory – weighed in on the debate.[31] It tested a scenario with carbon priced at a whopping $573 per ton, slapped on a resource extraction tax, and assumed rapid techno-logical innovation spurred by strong government support. The results? Resource use *still goes up*, nearly doubling by the middle of the century. As these results trickled out, UNEP had no choice but to change its position, admitting that green growth was a pipe dream: absolute decoupling of GDP and material use is simply not possible on a global scale.

What's going on here? What explains these bizarre results?

The thing about technology

Back in 1865, during the Industrial Revolution, the English economist William Stanley Jevons noticed something rather strange. James Watt had just introduced his steam engine, which was significantly more efficient than previous versions: it used less coal per unit of output. Everyone assumed that this would reduce total coal consumption. But oddly enough, exactly the opposite happened: coal consumption in England soared. The reason, Jevons discovered, was that the efficiency improvement saved money, and capitalists reinvested the savings to expand production. This led to economic growth – and as the economy grew, it chewed through more coal.

This odd result became known as the Jevons Paradox. In modern economics, the phenomenon is known as the Khazzoom-Brookes Postulate, named after the two economists who described it in the 1980s. And it doesn't just apply to energy – it applies to material resources too. When we innovate more efficient ways to use energy and resources, total consumption may briefly drop, but it quickly rebounds to an even higher rate. Why? Because companies use the savings to reinvest in ramping up more production. In the end, the sheer scale effect of growth swamps even the most spectacular efficiency improvements.[32]

Jevons described this as a 'paradox', but if you think about it it's not particularly surprising. Under capitalism, growth-oriented firms do not deploy new and more efficient technologies just for fun. They deploy them *in order to facilitate growth*. The same is true at the level of the whole economy. Ask any economist and they'll tell you: efficiency improvements are good because they stimulate economic growth. This is why we see that, despite constant improvements in efficiency, aggregate energy and resource use has been rising for the whole history of capitalism. There's no paradox; it's exactly what economists expect. Rising throughput happens not *despite* efficiency gains, but *because* of them. There's an important lesson here. The notion that continuous efficiency improvements will somehow magically lead to absolute decoupling is empirically and theoretically baseless.

But there's also something else going on. The technological innovations that have contributed most to growth have done so not because they enable us to use *less* nature, but because they enable us to use *more*.

Take the chainsaw, for instance. It's a remarkable invention that enables loggers to fell trees, say, ten times faster than they are able to do by hand. But logging companies equipped with chainsaws don't let their workers finish the job early and take the rest of the day off. They get them to cut down ten times as many trees as before. Lashed to the growth imperative, technology is used not to do the same amount of stuff in less time, but rather to do *more stuff in the same amount of time.*

The steam engine, the cotton gin, fishing trawlers – these technologies have contributed so spectacularly to growth not because money springs forth from them automatically, but because they have enabled capital to bring ever-greater swathes of nature into production. Innovations like containerisation and air freight contribute to growth because they enable goods to be transported from the point of extraction or production to the point of consumption more quickly. This even applies to seemingly immaterial innovations like Facebook's algorithms, which contribute to growth by allowing advertisers to get people to consume things they otherwise wouldn't. Facebook isn't a multi-billion-dollar company because it allows us to share pictures with each other, but because it expands the process of production and consumption.

Once we grasp how this works, it should come as no surprise that despite centuries of extraordinary innovation, energy and resource use keeps going up. In a system where technological innovation is leveraged to *expand* extraction and production, it makes little sense to hope that yet more technological innovation will somehow magically do the opposite.

There's a final problem. Scientists are beginning to realise that there are physical limits to how efficiently we can use resources. Sure, we might be able to produce cars and iPhones and skyscrapers more efficiently, but we can't produce them out of thin air. We might shift the economy to services such as yoga and movies, but even workout studios and cinemas require material inputs. There is always a limit to how 'lightweight' a product can be. And once we approach that limit, then continued growth causes resource use to start rising again.

This question was recently studied in detail by a team in Australia led by the scientist James Ward. They ran a series of models with extremely optimistic rates of technological innovation – well beyond what scientists consider to be feasible and faster than anything even green growth proponents have ever proposed. What they found is that while they were able to achieve some reductions in resource use in the short term, in the longer term resource use started rising again, *recoupling* with the rate of growth.

Ward's team say their findings constitute a 'robust rebuttal to the claim of absolute decoupling'. It is worth quoting their conclusion at length, as it has become quite famous in the field of ecological economics:

> We conclude that decoupling of GDP growth from resource use, whether relative or absolute, is at best only temporary. Permanent decoupling (absolute or relative) is

impossible for essential, non-substitutable resources because the efficiency gains are ultimately governed by physical limits. Growth in GDP ultimately cannot plausibly be decoupled from growth in material and energy use, demonstrating categorically that GDP growth cannot be sustained indefinitely. It is therefore misleading to develop growth-oriented policy around the expectation that decoupling is possible.

*

Let me be clear: technological innovation is absolutely important to the battle ahead. It is vital, in fact. We're going to need all the innovations and efficiency improvements we can get to drastically reduce the resource and carbon intensity of our economy. But the problem we face doesn't have to do with technology. The problem has to do with *growth*. Over and over again, we see that the growth imperative wipes out all the gains our best technology delivers.

We tend to think of capitalism as a system that incentivises innovation. And it does. But, paradoxically, the potential ecological benefits of innovation are constrained by the logic of capital itself. It doesn't have to be this way. If we lived in a different kind of economy – an economy not organised around growth – our technological innovations would have an opportunity to work as we expect them to. In a post-growth economy, efficiency improvements would actually reduce our impact on the planet. And once we are liberated from the growth imperative, we will be free to focus on different *kinds* of innovations – innovations designed to improve human and ecological welfare, rather than innovations designed to speed up the rate of extraction and production.

What about recycling?

There's another common fallacy that we need to face up to, and it has to do with recycling. The idea of a 'circular economy' has been gaining traction in policy circles recently as a response to the ecological crisis. These days everybody seems to be into it. The claim is that if we can scale up our recycling rate then we can keep growing GDP indefinitely, without worrying about the ecological impact of consumption. The European Union sees this as a plan to save capitalism, hoping that a circular economy will 'foster sustainable economic growth'.

Yes, we should absolutely aspire to a more circular economy. But the idea that recycling will save capitalism doesn't hold water. First, most of our material use cannot be recycled. Forty-four per cent of it is food and energy inputs, which become irreversibly degraded as we use them.[33] Twenty-seven per cent is *net addition* to stocks of buildings and infrastructure. Another big chunk is waste from mining.[34] In the end, only a small fraction of our total material use has circular potential. Even if we recycled all of it, economic growth would keep driving total resource use up. In any case, we're moving in the wrong direction: recycling rates have been *declining* over time, not improving. In 2018, the global economy achieved a recycling rate of 9.1%. Two years later it was down to 8.6%. This isn't because our recycling systems are getting worse. It's because growth in total material extraction and use is outstripping our gains in recycling. Once again, it's not our technology that's the problem – it's growth.[35]

But there's an even more fundamental problem with the idea of a 'green growth' circular economy. Even if we were able to recycle 100% of materials, that would pose a problem for the prospect of GDP growth. Growth tends to require an 'outside': an external source from which to extract value for free, or as close to free as possible. In a circular economy, the cost of materials is internalised. That's good from the perspective of ecology, but bad from the perspective of capital accumulation. Recycling costs money, and the cost of paying for recycled materials makes it more difficult to capture an ever-rising quantity of surplus. And the pinch gets tighter over time: materials degrade each time you recycle them, so you need ever-rising energy inputs – and ever-rising cost – in order to maintain their quality.

This same issue also poses a problem for those who say that all we need to do to solve the ecological crisis is put a 'price' on nature, and we can keep capitalism intact. If we could charge for 'ecosystem services' – say, the value added by earthworms and bees and mangroves – the market would respond accordingly and we'd be out of trouble. It's a nice thought, and recognising the value of nature would certainly be a step in the right direction. But remember, capitalist growth needs an 'outside'. To the extent that pricing nature would internalise the costs of production, it would pinch off prospects for growth. That's why no capitalist government has ever agreed to implement such a scheme in any serious way. In fact, it's the very reason we've failed for so long to get a decent price on carbon, which is effectively a price on nature. Internalising costs is important, but it is incompatible with the logic of capitalism.

Here's the bottom line: we should absolutely seek to build an economy that's as circular as possible! But the growth imperative makes this dream unnecessarily difficult to achieve. It would be much easier to improve circularity in a post-growth, post-capitalist economy.

The dystopia of green growth

The evidence piles up. And in the face of this evidence, proponents of green growth eventually begin to turn to fairy tales. Sure, they say, maybe green growth isn't empirically actual, but there's no reason that it can't happen *in theory*. We are limited only by our imagination! There's no reason we can't have our incomes rising for ever while we nonetheless consume less material stuff each year.

And here they are right. There's no *a priori* reason why such a thing can't happen in theory, in a magical alternative world. But there's a certain moral hazard at stake when we start trafficking in fairy tales – telling people not to worry because eventually, somehow, GDP will de-link from resource use and we'll be in the clear. In an era of climate emergency and mass extinction, we don't have time to speculate about imaginary possibilities. We don't have time to wait for this juggernaut of ecological destruction to suddenly stop being destructive, when all the evidence says it won't happen. It is unscientific, and a profoundly irresponsible gamble with human lives – with *all* of life.

There is an easy way to solve this problem. For decades, ecological economists have proposed that we can put an end to the debate once and for all with a simple and elegant intervention: impose a cap on annual resource use and waste, and tighten that cap year-on-year until we are back within planetary boundaries.[36] If green growthers really believe GDP will keep growing, for ever, despite rapid reductions in resource use, then this shouldn't worry them one bit. In fact, they should welcome such a move. It will give them a chance to prove to the world once and for all that they are right. Indeed, putting hard limits on resource use and waste will help incentivise the transition, spurring the shift toward dematerialised GDP growth.

But every time we propose this policy to green growthers, they wriggle away. Indeed, to my knowledge, not a single proponent of green growth has ever agreed to take it up. Why not? I suspect that on some deep level – despite the fairy tales – they realise that this is not how capitalism actually works. For 500 years, capitalist growth and accumulation has depended on extraction from nature. It has always needed an 'outside', external to itself, from which to plunder value, for free, without an equivalent return. To put a limit on resource extraction and waste is to effectively kill the goose that lays the golden eggs.

*

Let's pretend for a moment that they agree. Let's imagine that we cap resource use, scale it down to a sustainable level and hold it there. And let's imagine that the green growthers are correct, and GDP keeps growing by 3% a year, for ever. Remember, this is exponential, so in 200-some years global GDP is up to 1,000 times bigger than it is today. What will this hypothetical scenario look like, if efficiency improvements cannot deliver sufficient gains? When capital is no longer allowed to plunder nature in order to fuel growth and accumulation, we must ask ourselves: what new forms of exploitation might it devise?

The first culprit will be human labour. It's not difficult to imagine that if capital is rendered unable to exploit nature, it will double down instead on exploiting people. Capital already places extraordinary pressure on politicians around the world to cut wages and labour regulations. It's reasonable to expect that in a resource-cap scenario this pressure would intensify considerably. There would be a race to find more and ever-cheaper sources of labour.

But let's give the benefit of the doubt to green growthers, and assume that they're progressive enough to want to not only keep labour regulations but also improve them. Let's say we agree on an international minimum wage of some kind – a hard floor on labour exploitation to match the hard cap on material exploitation. In such a scenario, capital will be under enormous pressure to find new frontiers for surplus accumulation. It will need to find a 'fix' somewhere – new resources for appropriation, new outlets for investment, and new markets for sales. If surplus can't be extracted so easily from nature (because of the resource cap), and can't be extracted for free from humans (because of the wage floor), then where will it come from?[37]

Some economists say it will come from better products – products that are longer lasting, and higher quality. The products will be 'better' presumably because they embody more labour time, or more skill, or more advanced technology, and therefore they will be worth more money despite being made of less material. Here's the problem. Yes, we should absolutely strive for an economy focused on quality instead of quantity. But in order for this mechanism alone to drive growth at 3%, all products will have to be on average 3% 'better' per year, or 1,000 times better by the 2200s, and all this betterness would have to be reflected in a correspondingly higher cost. This would be strange for a few reasons.

First, if we think about the vast majority of things we need to live a good life, it's difficult to see how we would benefit much from them becoming 1,000 times better. A cancer treatment that is 1,000 times better, sure. But a table that is 1,000 times better? A hoodie that is 1,000 times better? Indeed, it begins to become absurd. Second, if products are 'better' because they are longer lasting, or more effective, this may well be *inimical* to growth, not conducive to it, as it reduces turnover. If our tables and hoodies last 1,000 times longer, then we will need to buy 1,000 times fewer of them. Third, if the betterness comes from more labour investment (say, a hand-woven hoodie rather than a mass-produced one), then we run into the problem of getting people to work much longer than before – which isn't ideal when the goal is to improve human lives.

Finally, in order for 'better' to translate into higher cost, the betterness has to be commodified (or enclosed). That might be OK in some cases, but in other cases we may want the opposite. For instance, if we develop better cancer treatments or other life-saving medicines, we may not want to charge people 1,000 times more to access them.

Let's not pretend either that capital's need for constant expansion is going to only make better products. That would be naïve. When capital has bumped up against limits to profit-growth in the past, it has found fixes in things like colonisation, structural adjustment programmes, wars, restrictive patent laws, nefarious debt instruments, land grabs, privatisation, and enclosing commons like water and seeds. Why would it be any different this time? Indeed, a study by the ecological economist Beth Stratford finds that when capital faces resource constraints, this is exactly what happens: it turns to aggressive rent-seeking behaviour.[38] It seeks to grab *existing* value wherever it can, with clever mechanisms to suck income and wealth from the public domain into private hands, and from the poor to the rich, exacerbating inequality.

Now, some might argue that capitalism could theoretically find growth opportunities in completely immaterial goods. That might sound nice on the face of it. But the thing about immaterial goods is that they tend already to be abundant and freely available, or are otherwise very easy to share. In order to secure growing profits in a context where all new value must be immaterial, then capital may well seek to enclose immaterial commons that are presently abundant and free, to make them artificially scarce and force people to pay for them. One can imagine an economy where not only water and seeds are privatised, commodifed and sold back to people for money, but also knowledge, songs and green spaces; maybe even parenting and physical touch; perhaps even the air itself. As for the rest of us, we would have to work more and more, producing (presumably) immaterial things for sale, simply in order to earn enough wages to buy access to immaterial things that we used to get for free.

The point here is that closing off the usual go-to fix (extraction from nature) will generate pressure for capital to find other fixes. That is the violent side of growth. It's naïve to pretend that these other fixes will somehow magically not be harmful, when we have 500 years of data to suggest that the reality is likely to be otherwise.

The unquestioned assumption

What is striking about all of this is that people are willing to go to extraordinary lengths to justify the continued pursuit of economic growth. Whenever there appears to be a conflict between ecology and growth, economists and politicians opt for the latter and try ever more creative ways to get reality to conform to it. Politicians are willing to bet everything on speculative technologies to avoid facing the imperative of radical emissions reductions. Proponents of green growth resort to outlandish imaginary scenarios and clever accounting tricks to maintain the illusion that we can carry on with the status quo. They are willing to risk everything – literally everything – just to keep GDP rising.

And yet, remarkably, none of these people has ever bothered to justify their core premise – the assumption that we *need* to keep increasing production, year-on-year, for ever. It is simply taken as an article of faith. Most people don't stop to question it, and indeed in some circles to do so is a kind of heresy. But what if this assumption is wrong? What if high-income countries *don't* need growth? What if we can improve human well-being without having to expand the economy at all? What if we can generate all the innovations we need for a rapid transition to renewable energy without a single dollar of additional GDP? What if instead of trying so desperately to decouple GDP from resources and energy use, we could decouple human progress from GDP instead? What if we could find a way to release our civilisation, and our planet, from the constraints of the growth imperative?

If we're willing to imagine speculative science-fiction fairy tales to keep the existing economy chugging along, why not just imagine a different kind of economy altogether?

PART TWO

Less is More

Four

Secrets of the Good Life

> Geese appear high over us,
> pass, and the sky closes. Abandon,
> as in love or sleep, holds
> them to their way, clear
> in the ancient faith: what we need
> is here.
>
> Wendell Berry

What explains the grip that growthism retains on our political imagination? We are told that no matter how rich a country becomes, its economy must keep growing, indefinitely, regardless of the costs. Economists and policymakers maintain this position even in the face of mounting evidence about ecological breakdown. When pressed, they offer a simple explanation: growth is responsible for the extraordinary improvements in welfare and life expectancy that we've witnessed over the past few centuries. We need to keep growing in order to keep improving people's lives. To abandon growth would be to abandon human progress itself.

It's a powerful narrative, and it seems so obviously correct. People's lives are clearly better now than they were in the past, and it seems reasonable to believe that we have growth to thank for that. But scientists and historians are now questioning this story. We have discovered that, remarkably for a claim that has become so entrenched in our society, the underlying empirical basis for it is weak. It turns out that the relationship between growth and human progress isn't quite as obvious as we once thought. It's not growth *itself* that matters – what matters is *what* we are producing, whether people have access to essential goods and services, and how income is distributed. And past a certain point, more GDP isn't necessary for improving human welfare at all.

Where does progress come from?

In the early 1970s, a British scholar called Thomas McKeown proposed a theory that would come to shape public narratives about growth for decades. McKeown was interested in historical trends in life expectancy. Looking at the data for Britain, he noticed that there was a striking rise in life expectancy after the 1870s – an improvement unlike anything else in the historical record. Like other scholars at the time, he was curious to know what had caused this apparently miraculous trend. It seemed a mystery. Most people assumed it had to do with innovations in modern medicine, which seems reasonable enough. But McKeown couldn't find much evidence for this. As he searched for an alternative theory, he landed on what seemed a sensible explanation: it must have been due to rising average incomes. After all, the Industrial Revolution was under way at that time, GDP was going up, and economic growth was making society richer. Surely this was the driving force behind health improvements.

McKeown's claims flipped the conventional wisdom upside down, and caused an immediate stir. At the same time, the American demographer Samuel Preston published another piece of evidence that seemed to bolster McKeown's thesis. Countries with higher GDP per capita also tend to have higher life expectancy. People in poor countries generally live shorter lives, while people in rich countries generally live longer lives. It's impossible to escape the obvious conclusion: GDP growth must be the key driver of progress on this core indicator of human welfare.

The McKeown Thesis and the Preston Curve, as they came to be known, captured the attention of economists and policymakers. This was at a time when the ideology of growthism was just beginning to take hold. It was the height of the Cold War, and the US government was peddling the idea that American-style capitalism was the world's ticket to 'development' and progress. McKeown's claims offered just the right evidence for this narrative, and the idea took off. Teams from the World Bank and the IMF went around the global South arguing that if governments wanted to improve social indicators like infant mortality and life expectancy, they needn't bother building public health systems (which many of them had been trying to do after the end of colonialism); instead, they should just focus on paving the way for growth. Do whatever it takes: get rid of environmental protections, slash labour laws, cut spending on healthcare and education, reduce taxes on the rich – it might seem regressive, and it may do a bit of harm in the short term, but ultimately it's the only true way to improve people's lives.

*

Those were heady days. During the 1980s and 1990s, the first two decades of the neoliberal era, this narrative reigned supreme. It served as the core justification for the structural adjustment programmes that were imposed so aggressively across the global South in the wake of the debt crisis. But research in the decades since has raised serious questions about the assumed equivalence between growth and human progress.

The thing is, when McKeown published his claims, he was not looking at long-term data. If he had been able to dig a bit more deeply into the historical record, he would have come to a rather different conclusion. As we saw in Chapter 1, the long rise of capitalism, from 1500 right into the Industrial Revolution, caused dramatic social dislocation everywhere it went. The enclosure movement in Europe, the Indigenous genocides, the Atlantic slave trade, the spread of European colonisation, the Indian famines; all of this took a measurable toll on human welfare around the world. The scars remain starkly visible in the public health record. For the vast majority of the history of capitalism, growth didn't deliver welfare improvements in the lives of ordinary people; in fact, it did exactly the opposite.[1] Remember, capitalist expansion relied on the creation of artificial scarcity. Capitalists enclosed the commons – lands, forests, pastures and other resources that people depended on for survival – and ripped up subsistence economies in order to push people into the labour market. The threat of hunger was used as a weapon to enforce competitive productivity. Artificial scarcity quite often caused the livelihoods and welfare of ordinary people to collapse even as GDP grew.

It wasn't until nearly 400 years later that life expectancies in Britain finally began to recover, unleashing the rising trend that McKeown had noticed. It happened slightly later in the rest of Europe, while in the colonised world longevity didn't begin to improve until the early 1900s. So if growth itself does not have an automatic relationship with life expectancy and human welfare, what could possibly explain this trend?

Historians today point out that it began with a startlingly simple intervention, something McKeown had overlooked: sanitation.[2] In the middle of the 1800s, public health researchers had discovered that health outcomes could be improved by introducing simple sanitation measures, such as separating sewage from drinking water. All it required was a bit of public plumbing. But public plumbing requires public works, and public money. You have to appropriate private land for things like public water pumps and public baths. And you have to be able to dig on private property in order to connect tenements and factories to the system. This is where the problems began. For decades, progress towards the goal of public sanitation was opposed, not enabled, by the capitalist class. Libertarian-minded landowners refused to allow officials to use their property, and refused to pay the taxes required to get it done.

The resistance of these elites was broken only once commoners won the right to vote and workers organised into unions. Over the following decades these movements, which in Britain began with the Chartists and the Municipal Socialists, leveraged the state to intervene against the capitalist class. They fought for a new vision: that cities should be managed for the good of everyone, not just for the few. These movements delivered not only public sanitation systems but also, in the years that followed, public healthcare, vaccination coverage, public education, public housing, better wages and safer working conditions. According to research by the historian Simon Szreter, access to these public goods – which were, in a way, a new kind of commons – had a significant positive impact on human health, and spurred soaring life expectancy through the twentieth century.[3]

This explanation is now backed up by a strong consensus among public health researchers. Empirical data from the United States shows that water sanitation measures alone explain three quarters of the decline in infant mortality in major cities between 1900 and 1936, and nearly half the decline in total mortality.[4] A recent study led by an international team of medical scientists found that, after sanitation, the greatest predictor of improved life expectancy is access to universal healthcare, including child vaccination.[5] And once you have these basic interventions in place, the biggest single driver of continued improvements in life expectancy happens to be education – and particularly women's education. The more you learn, the longer you live.[6]

None of this should be particularly surprising. Those who bothered to read Samuel Preston's original paper in 1975 would have noticed that he himself observed that up to 90% of improvements in global life expectancy between the 1930s and the 1960s were attributable to factors 'exogenous to income', such as public health programmes and other social technologies.[7] Forty years later, in 2015, the UN Development Programme published an analysis confirming that the relationship between economic growth and changes in health and education is 'weak', and concluding that 'human development is different from economic growth.'[8]

Don't get me wrong. It's true that nations with higher income tend in general to have better life expectancies than nations with lower income. But there is no simple *causal* relationship between these two variables. 'The historical record is clear that economic growth itself has no direct, necessary positive implications for population health,' Szreter points out. 'The most that can be said is that it creates the longer-term *potential* for population health improvements.'[9] Whether or not that potential is realised depends on the political forces that determine what kinds of things get produced, who has access to them, and how income is distributed. Progress in human welfare has been driven by progressive political movements and governments that have managed to harness resources to deliver robust public goods and fair wages.[10] In fact, the historical record shows that in the absence of these forces, growth has quite often worked *against* social progress, not for it.

Reclaiming the commons

Of course, things like universal healthcare, sanitation, education and decent wages require mobilising resources. Economic growth can absolutely help towards that end, and in poor countries it is even necessary. But – and here's the crucial bit – the interventions that really matter when it comes to improving human welfare do not require high levels of GDP. The relationship between GDP and human welfare plays out on a saturation curve, with sharply diminishing returns: after a certain point, which high-income nations have long surpassed, more GDP does little to improve core social outcomes.[11] The relationship begins to break down.

In fact, there are many countries that manage to achieve strikingly high levels of human welfare with relatively little GDP per capita. We tend to see these countries as 'outliers', but they prove the very point that Szreter and other public health researchers have been trying to make: it's all about distribution. And what matters most of all is investment in universal public goods. This is where things get interesting.

Take life expectancy, for example. The United States has a GDP per capita of $59,500, making it one of the world's richest countries. People in the US can expect to live 78.7 years, nudging them just into the top 20%. But dozens of countries beat the US on this crucial indicator with only a fraction of the income. Japan has 35% less income than the US, but a life expectancy of 84 years – the highest in the world. South Korea has 50% less income and a life expectancy of 82 years. And then there's Portugal, which has 65% less income and a life expectancy of 81.1 years. This is not a matter of just a few special cases. The European Union as a whole has 36% less income than the United

States, and yet beats the US not only on life expectancy but on virtually every other indicator of human welfare.

Then there's Costa Rica, which provides perhaps the most astonishing example. The rainforest-rich Central American country beats the US on life expectancy despite having 80% less income. Indeed, Costa Rica ranks among the most ecologically efficient economies on the planet, in terms of its ability to deliver high standards of welfare with minimal pressure on the environment. And when we look at it across time, the story becomes even more fascinating: Costa Rica managed to achieve some of its most impressive gains in life expectancy during the 1980s, catching up to and surpassing the United States, during a time when its GDP per capita was not only small (one-seventh that of the US) but *not growing at all.*

It's not just life expectancy that behaves like this. We can see the same pattern playing out when it comes to education. Finland is widely known as having one of the best education systems in the world, despite having a GDP per capita that's 25% less than the United States. Estonia is right towards the top of world education rankings too, but with 66% less income than the US.[12] Poland outperforms the US with 77% less. On the UN's education index, the nation of Belarus beats high-performers like Austria, Spain, Italy and Hong Kong with a GDP per capita that's a full 90% less than the US.

What explains the remarkable results that these countries have achieved? It's simple: they've all invested in building high-quality universal healthcare and education systems.[13] When it comes to delivering long, healthy, flourishing lives for all, this is what counts.

The good news is that this is not at all expensive to do. In fact, universal public services are significantly more cost-effective to run than their private counterparts. Take Spain, for example. Spain spends only $2,300 per person to deliver high-quality healthcare to everyone as a fundamental right, achieving one of the highest life expectancies in the world: 83.5 years; a full five years longer than Americans. By contrast, the private, for-profit system in the United States sucks up an eye-watering $9,500 per person, while delivering lower life expectancy and worse health outcomes.

There are similarly promising examples emerging in pockets across the global South. Countries whose governments have invested in universal public healthcare and education have seen some of the world's fastest improvements in life expectancy and other indicators of human welfare.[14] Sri Lanka, Rwanda, Thailand, China, Cuba, Bangladesh and the Indian state of Kerala – all are achieving astonishing gains despite having relatively low GDP per capita. Indeed, several studies have found that countries with universal public provisioning systems are able to achieve better social outcomes than countries without such systems, at any given level of economic development.[15]

Over and over again, the empirical evidence shows that it is possible to achieve high levels of human development *without* high levels of GDP. According to data from the UN, nations can reach the very highest category on the life expectancy index with as little as $8,000 per capita (in terms of purchasing power parity, or PPP), and very high levels on the education index with as little as $8,700. In fact, nations can succeed on a wide range of key social indicators – not just health and education, but also employment, nutrition, social support, democracy and life satisfaction – with less than $10,000 per capita, while staying within or near planetary boundaries.[16] What's remarkable about these figures is that they are well below the world average GDP per capita $17,600 PPP. In other words, in theory we could achieve all of our social goals, for every person in the world, with *less* GDP than we presently have, simply by organising production around human well-being, investing in public goods, and distributing income and opportunity more fairly.

So it's clear that the relationship between GDP and human welfare breaks down after a certain point. But there is also something else interesting about this relationship. Past a certain threshold, more growth actually begins to have a *negative* impact. We can see this effect when we look at alternative metrics of progress, like the Genuine Progress Indicator. GPI starts with personal consumption expenditure (which is also the starting point for GDP) and adjusts for income inequality as well as the social and environmental costs of economic activity. By accounting for the costs as well as the benefits of growth, this measure gives us a more balanced view of what's happening in an economy. When we plot this data over time, we see that global GPI grew along with GDP until the mid-1970s, but since then has flattened out and even declined, as the social and environmental costs of growth have become significant enough to cancel out consumption-related gains.[17]

As the ecologist Herman Daly has put it, after a certain point growth begins to become 'uneconomic': it begins to create more 'illth' than wealth. We can see this happening on a number of fronts: the continued pursuit of growth in high-income nations is exacerbating inequality and political instability,[18] and contributing to problems like stress and depression from overwork and lack of sleep, ill health from pollution, diabetes and heart disease, and so on.

*

I was blown away when I first learned about all this. It is powerful because it enables us to think about growth a bit differently. From the perspective of human welfare, the high levels of GDP that characterise the United States, Britain and other higher-income countries turn out to be vastly in excess of what they actually need.

Consider this thought experiment: if Portugal has higher levels of human welfare than the United States with $38,000 less GDP per capita, then we can conclude that $38,000 of America's per capita income is effectively 'wasted'. That adds up to $13 trillion per year for the US economy as a whole. That's $13 trillion worth of extraction and production and consumption each year, and $13 trillion worth of ecological pressure, that adds nothing, in and of itself, to the fundamentals of human welfare. It is damage without gain. This means that the US economy could in theory be scaled down by a staggering 65% from its present size while at the same time *improving* the lives of ordinary Americans, if income was distributed more fairly and invested in public goods.

Of course, we might expect that some of the excess income and consumption we see in rich countries yield improvements in quality of life that are not captured by data on life expectancy and education. What about things like happiness and well-being? Surely as GDP goes up, these more subjective indicators will rise too? It seems like a reasonable assumption; after all, the American Dream promises that income and consumption is the ticket to happiness. But strangely enough, when we look at measures of overall happiness and well-being, it turns out that even these indicators have a tenuous relationship with GDP. This rather puzzling result is known as the Easterlin Paradox, after the economist who first pointed it out.

In the United States, happiness rates peaked in the 1950s, when GDP per capita was only about $15,000 (in today's dollars). Since then the average real income of Americans has quadrupled, and yet happiness has plateaued and even declined for the past half-century. The same is true of Britain, where happiness has declined since the 1950s despite a tripling of income.[19] Similar trends are playing out in country after country.

What explains this paradox? Researchers have found that – once again – it's not income itself that matters, but how it's distributed.[20] Societies with unequal income distribution tend to be less happy. There are a number of reasons for this. Inequality creates a sense of unfairness; it erodes social trust, cohesion and solidarity. It's also linked to poorer health, higher levels of crime and less social mobility. People who live in unequal societies tend to be more frustrated, anxious, insecure and discontent with their lives. They have higher rates of depression and addiction.

It's easy to imagine how this might play out in real life. If you get a raise at work it's bound to boost your happiness. But what happens when you discover that your colleagues got a raise that was twice what you received? Suddenly you're not happy at all – you're upset. You feel devalued. Your sense of trust in your boss takes a hit, and your sense of solidarity with your colleagues falls apart.

Something similar happens when it comes to consumption. Inequality makes people feel that the material goods they have are inadequate. We constantly want more, not because we need it but because we want to keep up with the Joneses. The more our friends and neighbours have, the more we feel that we need to match them just to feel like we're doing OK. The data on this is clear: people who live in highly unequal societies are more likely to shop for luxury brands than people who live in more equal societies.[21] We keep buying more stuff in order to feel better about ourselves, but it never works because the benchmark against which we measure the good life is pushed perpetually out of reach by the rich (and, these days, by social media influencers). We find ourselves spinning in place on an exhausting treadmill of needless over-consumption.

So, if not income, what *does* improve well-being? In 2014, the political scientist Adam Okulicz-Kozaryn conducted a review of existing data on this question. He found something remarkable: countries that have robust welfare systems have the highest levels of human happiness, when controlling for other factors. And the more generous and universal the welfare system, the happier everyone becomes.[22] This means things like universal healthcare, unemployment insurance, pensions, paid holiday and sick leave, affordable housing, daycare and strong minimum wages. When people live in a fair, caring society, where everyone has equal access to social goods, they don't have to spend their time worrying about how to cover their basic needs day to day – they can enjoy the art of living. And instead of feeling they are in constant competition with their neighbours, they can build bonds of social solidarity.

This explains why there are so many countries that have higher levels of well-being than the United States, even with significantly less GDP per capita. It's a long list that includes Germany, Austria, Sweden, the Netherlands, Australia, Finland, Canada and Denmark – the classic social democracies. But it also includes Costa Rica, which matches the United States on well-being indicators with only a fifth of the income.[23] In all of these cases, their success is down to strong social provisions.

The data on happiness is remarkable. But some researchers have pointed out that we shouldn't be satisfied with looking just at happiness. We should look at people's sense of *meaning* – a more profound state that lies beneath the tumult of daily emotions.

And when it comes to meaning, what matters has even less to do with GDP. People feel they have meaningful lives when they have the opportunity to express compassion, co-operation, community and human connection. These are what psychologists refer to as 'intrinsic values'. These values don't have to do with external indicators like how much money you have, or how big your house is; they run much deeper than that. Intrinsic values are far more powerful, and more durable, than the fleeting rush we might get from a boost in income or material consumption.[24] We humans are evolved for sharing, co-operation and community. We flourish in contexts that enable us to express these values, and we suffer in contexts that stifle them.

Meaning has a real, material impact on people's lives. In 2012, a team of researchers from Stanford School of Medicine visited the Nicoya Peninsula in Costa Rica to try to make sense of some fascinating data coming out of that region. We know that Costa Ricans live long lives: around eighty years on average. But the researchers had noticed that Nicoyans live even longer, with a life expectancy of up to eighty-five years – one of the highest in the world. This is odd, because Nicoya is one of the poorest parts of Costa Rica, in monetary terms. It is a subsistence economy where people live traditional, agricultural lifestyles. So what explains these results? Costa Rica has an excellent public health-care system, so that's a big part of it. But the researchers found that Nicoyans' extra longevity is due to something more. Not diet, not genes, but something completely unexpected: community. The longest-living Nicoyans all have strong relationships with their families, friends and neighbours. Even in old age, they feel connected. They feel valued. In fact, the poorest households have the longest life expectancies, because they are more likely to live together and rely on each other for support.[25]

Imagine. People living subsistence lifestyles in rural Costa Rica enjoy longer, healthier lives than people in the richest economies on Earth. North America and Europe may have highways and skyscrapers and shopping malls, huge homes and cars and glitzy institutions – all the markers of 'development'. And yet none of this gives them a shred of advantage over the fishermen and farmers of Nicoya when it comes to this core measure of human progress. The data piles up. Over and over again, we see that the excess GDP that characterises the richest nations wins them nothing when it comes to what really matters.

Flourishing without growth

All this amounts to excellent news. It means that upper-middle income and high-income nations can deliver good lives for all, achieving real progress in human development, without needing growth in order to do so. We know exactly what works: reduce inequality, invest in universal public goods, and distribute income and opportunity more fairly.

What's exciting about this approach is that it also has a direct positive impact on the living world. As societies become more egalitarian, people feel less pressure to pursue ever-higher incomes and more glamorous status goods. This liberates people from the treadmill of perpetual consumerism. Take Denmark, for example. Consumer research shows that because Denmark is more equal than most other high-income countries, people buy fewer clothes – and keep them for longer – than their counterparts elsewhere. And firms spend less money on advertising, because people just aren't as interested in unnecessary luxury purchases.[26] This is one of the reasons why more egalitarian societies turn out to have lower levels of per capita emissions, when correcting for other factors.[27]

But reducing inequality reduces ecological impact in more direct ways too. Rich people have a much higher ecological footprint than everybody else. The richest 10% of the world's population are responsible for more than half of the world's total carbon emissions since 1990. In other words, the global climate crisis is being driven largely by the global rich. And things become even more lopsided as we climb the income ladder. Individuals in the richest 1% emit one hundred times more than individuals in the poorest half of the human population.[28] Why? It's not only because they consume more stuff than everybody else, but also because the stuff they consume is more energy-intensive: huge houses, big cars, private jets, frequent flights, long-distance holidays, luxury imports, and so on.[29] And if the rich have more money than they can spend, which is virtually always the case, then they invest their excess in expansionary industries that are quite often ecologically destructive.

This leads us to a simple but radical conclusion: any policy that reduces the incomes of the very rich will have a positive ecological benefit. And because the excess incomes of the rich win them nothing when it comes to welfare, this can be accomplished without any cost to social outcomes. This position is widely shared among researchers who study this issue. The French economist Thomas Piketty, one of the world's leading experts on inequality, doesn't mince his words: 'A drastic reduction in purchasing power of the richest would therefore in itself have a substantial impact on the reduction of emissions at global level.'[30]

There are also ecological benefits to be reaped from investing in public services. Public services are almost always less intensive than their private equivalents. Britain's National Health Service, for instance, emits only one-third as much CO_2 as the American health system, and delivers better health outcomes in the process. Public transportation is less intensive in terms of both energy and materials than private cars. Tap water is less intensive than bottled water. And things like public parks, swimming pools and recreational facilities are less intensive than everyone buying bigger yards, private pools and personal gym equipment. Plus, it's more fun. If you visit Finland, you'll find a whole society that thrives on the conviviality of public saunas – it's a national pastime that plays a big role in making Finland one of the happiest countries in the world.[31]

Shared public goods also take pressure off people's need for private income. Take the United States, for example. Americans are under extraordinary pressure to work ever-longer hours and pursue ever-higher incomes, because the cost of accessing basic goods like healthcare and education is not only outrageously high, but constantly rising. Decent health insurance can be prohibitively expensive to buy, and the cost of deductibles and co-payments is often enough to sink people into debt for their whole lives. Health insurance premiums have nearly quadrupled since 2000.[32] As for education, a family with two kids can expect to pay up to half a million dollars just to put them through college – almost 500% more than in the 1980s.[33] These prices have nothing to do with the 'real' cost of healthcare and education: they are an artefact of a system organised around profit.

Now, consider this: if the US were to transition to a public health and education system, people would be able to access the goods they need to live well for a mere fraction of the cost. Suddenly they would be under much less pressure to pursue high incomes just in order to get by.

*

This brings us to the crucial point. When it comes to human welfare, it's not income *as such* that matters. It's what that income can buy, in terms of access to the things we need to live well. It's the 'welfare purchasing power' of income that counts. Trying to run a household on $30,000 in the United States would be a struggle. You can forget sending your kids to a decent university. But the exact same income in Finland, where people enjoy universal healthcare and education and rent controls, would feel luxurious. By expanding people's access to public services and other commons, we can improve the welfare purchasing power of people's incomes, enabling flourishing lives for all without needing any additional growth. Justice is the antidote to the growth imperative – and key to solving the climate crisis.

This means fundamentally reversing the neoliberal policies that have dominated for the past forty years. In their desperate hunt for growth, governments have privatised public services, slashed social spending, cut wages and labour protections, handed tax cuts to the richest and sent inequality soaring. In an age of climate breakdown, we need to be doing exactly the opposite.

The weight of evidence is clear that we do not need more growth in order to achieve our social goals. And yet growthist narratives nonetheless have remarkable staying power. Why? Because growth serves the interests of the richest and most powerful factions in our society. Take the United States, for example. Real GDP per capita in the US has doubled since the 1970s. One might assume that such extraordinary growth would have delivered decisive improvements to human lives, but in fact the opposite has happened. The poverty rate today is higher, and real wages are lower, than they were forty years ago.[34] Despite half a century of growth the country has *regressed* on these core indicators, while virtually all of the gains have gone to the already-rich. The annual incomes of the richest 1% have more than tripled over this period, soaring to an average of $1.4 million *per person*.[35]

With data like this on the table, it becomes clear that growthism is little more than ideology – an ideology that benefits a few at the expense of our collective future. We're all pushed to step on the accelerator of growth, with deadly consequences for our living planet, all so that a rich elite can get even richer. From the perspective of human life, this is clearly an injustice. And indeed we have been aware of this problem for some time. But from the perspective of ecology, it is even worse – it is a kind of madness.

Justice for the South

Richer countries don't need growth in order to improve people's lives. But what about poorer countries? Take the Philippines, for example. These islands in the western Pacific fall short on a number of key social indicators, including life expectancy, sanitation, nutrition and income. But they also remain well within safe planetary boundaries, in terms of their use of land, water, energy, material resources and so on.[36] There's no reason that the Philippines shouldn't increase its use of these resources to the extent that doing so is necessary to meet people's needs. The same is true for most countries in the global South.

Here's the good news. My colleagues and I have analysed data for over 150 nations, and our results show that it's possible for global South countries to achieve strong outcomes on key human development indicators (including life expectancy, well-being, sanitation, education, electricity, employment and democracy) while remaining within or near planetary boundaries. Here again, Costa Rica – which I described above – provides an excellent example of what this might look like.[37] Other research has demonstrated that it is possible to deliver good lives for all – including universal healthcare, education, housing, electricity, heating/cooling, public transit, computing and so on – with levels of energy use that are compatible with keeping global warming to less than 1.5 degrees.[38] But this requires an entirely different way of thinking about development. Instead of pursuing growth for its own sake and hoping that it will magically improve people's lives, the goal must be to focus on improving people's lives first and foremost – and if that requires or entails economic growth, then so be it. In other words, organise the economy around the needs of humans and ecology, rather than the other way around.

This approach to development has a long history in the global South. It was championed by anti-colonial leaders including Mahatma Gandhi, Patrice Lumumba, Salvador Allende, Julius Nyerere, Thomas Sankara and dozens of other figures who insisted on a human-centred economics, with an emphasis on the principles of justice, well-being and self-sufficiency. But perhaps no one from that era expressed these ideas more succinctly than Frantz Fanon, the revolutionary intellectual from Martinique, when in the 1960s he penned these words that I think continue to resonate today:

> Come, then, comrades, the European game has finally ended; we must find something different. We today can do everything, so long as we do not imitate Europe, so long as we are not obsessed by the desire to catch up with Europe. Europe now lives at such a mad, reckless pace that she has shaken off all guidance and all reason, and she is running headlong into the abyss; we would do well to avoid it with all possible speed. The Third World today faces Europe like a colossal mass whose aim should be to try to resolve the problems to which Europe has not been able to find the answers. But let us be clear: what matters is to stop talking about output, and intensification, and the rhythm of work. No, we do not want to catch up with anyone. What we want to do is to go forward all the time, night and day, in the company of Man, in the company of all men. So, comrades, let us not pay tribute to Europe by creating states, institutions and societies which draw their inspiration from her. Humanity is waiting for something other from us than such an imitation.[39]

What Fanon is calling for here is a kind of decolonisation – that we should decolonise the imaginary of economic development and allow different approaches to flourish.[40] What does this look like, in practice? It means, following the example of states like Costa Rica, Sri Lanka, Cuba and Kerala, investing in robust universal social policy to guarantee healthcare, education, water, housing, social security. It means land reform so that small farmers have access to the resources that they need to thrive. It means using tariffs and subsidies to protect and encourage domestic industries. It means decent wages, labour laws and a progressive distribution of national income. And it means building economies that are organised around renewable energy and ecological regeneration rather than around fossil fuels and extractivism.

It's important to remember that many of these policies were used widely across the South in the post-colonial decades, from the 1950s to the 1970s, before that vision was dismantled by structural adjustment programmes from the 1980s onwards. A few countries managed to escape this fate, for various historical reasons. Costa Rica was one of them. So were South Korea and Taiwan (although their ecological policies have fallen short). They continued to pursue a more progressive approach to economic policy, and continued investing in public services – and today they enjoy high levels of human development as a result. They stand as beacons of what the South could have achieved, had it been left alone.

What the South needs, then, is to be free from structural adjustment – in other words, free from control by foreign creditors – so governments can pursue the progressive economic policies that we know to be so effective at delivering human development. And this brings me to an important point: when it comes to progress in the South, this is about more than just domestic policy – it's about global justice.

*

When people think of the global poor, they quite often imagine people living in countries that are cut off from the world economy – backwaters untouched by globalisation, and unconnected from the lives of people in rich countries. But this image gets it completely wrong. The poor are deeply integrated into the circuits of global capital. They work in sweatshops for multinational companies like Nike and Primark. They risk their lives mining the rare-earth minerals that we depend on for our smartphones and computers. They harvest the tea leaves and coffee beans and sugar cane that most people consume every day. They pick the berries and bananas that Europeans and North Americans eat every morning for breakfast. And in many cases theirs is the land from which the oil and coal and gas that power the global economy is extracted – or at least it used to be, before it was taken from them. All told, they contribute the vast majority of the labour and resources that go into the global economy.[41]

And yet in return for this they receive literally pennies. The poorest 60% of humanity receives only about 5% of total global income.[42] Over the course of the past four decades since 1980, their daily incomes have increased by an average of about 3 cents per year.[43] Forget 'trickle-down' economics – this is barely even a vapour.

For the world's rich, by contrast, it's a rather different story. Over the four decades since 1980, no less than 46% of all new income from global economic growth has gone to the richest 5%. The richest 1% alone capture $19 trillion in income every year, which represents nearly a quarter of global GDP.[44] That adds up to more than the GDP of 169 countries *combined* – a list that includes Norway, Sweden, Switzerland, Argentina, all of the Middle East and the entire continent of Africa. The rich lay claim to an almost unimaginable share of the income the global economy generates; income that is extracted from the lands and bodies of the poor.

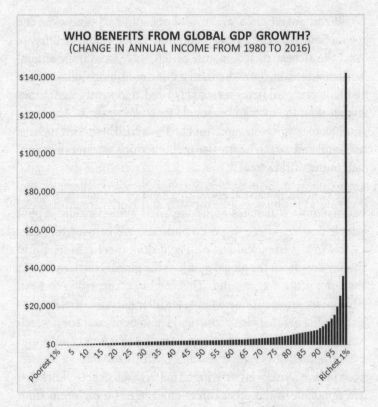

WHO BENEFITS FROM GLOBAL GDP GROWTH?
(CHANGE IN ANNUAL INCOME FROM 1980 TO 2016)

This graph shows the average gains in income for individuals within each percentile. Source: World Inequality Database (Constant 2017 USD). Data management by Huzaifa Zoomkawala.

To put these sums in perspective, consider this: to bring everyone in the world above the income poverty line of $7.40 per day, and to provide universal public healthcare for every person in the global South at a standard equivalent to that in Costa Rica, would require about $10 trillion.[45] That's a significant sum, on the face of it. But notice that it's only half of the annual income of the richest 1%. In other words, if we were to shift $10 trillion of excess annual income from the richest 1% to the global poor, we could end poverty in a stroke, and boost life expectancy in the global South to eighty years – eliminating the global health gap. And the richest 1% would still be left with an average annual household income of more than a quarter million dollars: more than anyone could ever reasonably need, and nearly eight times higher than the median household income in Britain. And that's just income; we haven't even touched wealth. The richest 1% have accumulated wealth worth $158 trillion, which amounts to nearly half of the world's total.[46]

There is nothing natural about this kind of inequality. It exists because powerful states and companies systematically exploit people and resources in poor countries. We can see this very clearly in the empirical record. Right now, more resources and money flow from the global South to the global North each year than the other way around. This might be surprising to hear, because we are accustomed to the familiar narrative that empha-sises all the aid that rich countries give to poor countries, which amounts to around $130 billion a year. But that flow of aid – and even the flow of private investment, which adds up to another $500 billion a year – is outstripped many times over by flows that are siphoned in the other direction. There is a *net drain* from poor countries to rich countries.

Once we grasp these facts, it becomes clear that achieving devel-opment in the South is a matter of ending patterns of extraction and exploitation, and changing the rules of the economy to make it fundamentally fairer for the world's majority. In my last book, *The Divide*, I explored what this might look like. Instead of repeating that work here, let me offer just a few brief examples.

Take labour, for instance. We know that growth in the global North depends in large part on the labour of workers in the South. But researchers estimate that people who work in export industries in the South lose around $2.8 trillion in underpaid wages each year, because they lack bargaining power in international trade.[47] One straightforward way to address this problem would be to introduce a global minimum wage. It could be managed by the International Labour Organization, and either fixed as a percentage of each country's median income or set at local living-income thresholds.

Then there are illicit financial flows. Right now, some $1 trillion is stolen out of global South countries each year and stashed in offshore secrecy jurisdictions, mostly by multinational companies seeking to evade taxes.[48] For example, companies might generate profits in Guatemala or South Africa, but then shift that money into tax havens like Luxembourg or the British Virgin Islands. This starves global South countries of the revenues they need to invest in public services. But it is not an intractable problem: we could shut down the tax evasion system with laws to regulate cross-border trade and corporate accounting.

Another problem is that the international institutions that govern the global economy are deeply anti-democratic, and tilted heavily in favour of rich nations. At the World Bank and the IMF, the United States holds veto power over all major decisions, and high-income countries control the majority of the vote. In the World Trade Organization, bargaining power depends largely on GDP, so the countries that grew rich during the colonial period get to determine the rules of international trade. Democratising these institutions would ensure that global South countries have a real say in the decisions that affect them, and greater control over their own economic policy. The UN estimates that fairer trade rules at the WTO could allow poor countries to earn over $1.5 trillion in additional export revenues each year.[49]

There are many other interventions we might consider. We could cancel odious debts, freeing poor countries to invest in public healthcare and education instead of spending all their money on interest payments to foreign banks; we could put an end to corporate land grabs, and distribute land back to small farmers; we could reform the subsidy regimes that give high-income countries an unfair advantage in the agricultural industry. Changes like these would enable the people of the South to capture a fairer share of income from the global economy, and access the resources they need to ensure good lives for all.

Breaking free of ideology

Once we grasp the scale of national and global inequalities, then the narrative that seeks to cast GDP growth as a proxy for human progress begins to seem a bit tendentious – perhaps even a bit ideological. And by ideology I mean in the technical sense: a set of ideas promoted by the dominant class, which serves their material interests, and which everybody else has internalised to such an extent that they are willing to go along with a system they might otherwise reject as unjust. The Italian philosopher Antonio Gramsci has called this 'cultural hegemony': when an ideology becomes so normalised that it is difficult or even impossible to reflect on it.

The elites of our world know very well what's going on here. It would be silly to assume they don't. They know the data on income distribution. They live by that data. They spend their lives thinking of ways to increase their share of national and global income. Their call for more growth is ultimately about speeding up the mechanisms of accumulation; the claims about the putative relationship between growth and human progress are just an alibi. Of course, they hope that growth will end up improving the incomes of the poor, and in so doing pacify social conflict. After all, elite accumulation is more politically palatable if the incomes of the poor are rising too. But this strategy cannot be sustained in an era of ecological crisis. Something has to give.

The problem with growthism is that for decades it has distracted us from the difficult politics of distribution. We have ceded our political agency to the lazy calculus of growth – the notion that growth is automatically good for everyone. The climate emergency changes this. It forces us to face up to the brutal inequalities of the global economy. It forces us into the zone of political contestation. The notion that we need *aggregate* growth to improve people's lives no longer makes any sense. We need to be able to specify growth for whom, and for what ends. We must learn to ask: where does the money go? Who benefits from it? In an era of ecological breakdown, are we really content to accept an economy where nearly a quarter of total output goes into the pockets of millionaires?

Henry Wallich, a former member of the US Federal Reserve Board, once famously pointed out that 'growth is a substitute for equality of income'. And it's true: it is politically easier to rev up GDP and hope some of it trickles down to the poor than it is to distribute existing income more fairly. But we can flip Wallich's logic around: if growth is a substitute for equality, then equality can be a substitute for growth.[50] We live on an abundant planet. If we can find ways to share what we already have more fairly, we won't need to plunder the Earth for more. Justice is the antidote to growth.

Those who insist that aggregate growth is necessary to improve people's lives force us into a horrible double-bind. We are made to choose between human welfare or ecological stability – an impossible choice that nobody wants to face. But when we understand how inequality works, suddenly the choice becomes much easier: between living in a more equitable society, on the one hand, and risking ecological catastrophe on the other. Most people would have little difficulty choosing. Of course, achieving this will not be easy. It will require an enormous struggle against those who benefit so prodigiously from the status quo. And presumably this is why some are so eager that we avoid this course of action: they would prefer to sacrifice the planet in order to maintain the existing distribution of global income.

What about innovation?

There's another powerful narrative that we need to grapple with. The dominant story holds that growth isn't just necessary for *human* progress, it's also necessary for *technological* progress. Most pressingly, growth is the only way to mobilise the financial resources for the energy transition, and the only way to get the innovation we need in order to make our economies more efficient.

There's no question we need innovation to solve the climate crisis. We need better solar panels, better wind turbines, better batteries, and we need to figure out how to dismantle the global fossil fuel infrastructure and replace it with renewables. That's a big challenge. But here's the good news: we don't need growth in order to do it.

First, there is no evidence to back up the assumption that *aggregate* growth is necessary for achieving these goals. It doesn't make sense to grow the whole GDP and just blindly hope that it will magically end up invested in solar panel factories. If that's how the Allies had approached the need for tanks and aircraft during the Second World War, the Nazis would be in charge of Europe right now. This kind of mobilisation requires government policy to guide and direct *existing* financial resources. The vast majority of major, collaborative infrastructure projects around the world have been guided by government policy and funded by public resources: sanitation systems, road systems, railway networks, public health systems, national power grids, the postal service. These are not the spontaneous outcomes of market forces, much less of abstract growth. Projects like these require public investment. Once we realise this, it becomes clear that we can fund the transition quite easily by directing existing public resources from, say, fossil fuel subsidies (which presently stand at \$5.2 trillion, 6.5% of global GDP) and military expenditure (\$1.8 trillion) into solar panels, batteries and wind turbines.[51]

Government policy can also be used to guide *private* investment. We know that when governments begin investing in specific sectors it 'crowds in' other investors who are eager to take advantage of incentives or provide necessary upstream supplies.[52] On top of this, simple rules can be introduced that require large companies and rich individuals to use a share of their income (say, 5%) to buy bonds designed to fund specific projects – like a rapid rollout of renewable energy. Such measures have been used by governments many times in the past – such as during the New Deal in the United States, and during the developmentalist period in the global South – and there's no reason we can't do it again.

As for the process of innovation itself: it's important to remember that many of the most important innovations of the modern era, including truly life-changing technologies we use every day, were funded not by growth-oriented firms but rather by public bodies. From plumbing to the internet, vaccines to microchips, even the technologies that make up smartphones – all of these came from publicly funded research. We don't need *aggregate* growth to deliver innovation. If the objective is to achieve specific kinds of innovation, then it makes more sense to invest in those directly, or incentivise investment with targeted policy measures, rather than grow the whole economy indiscriminately and hope it will deliver the innovation we want. Is it really reasonable to grow the plastics industry, the timber industry and the advertising industry in order to get more efficient trains? Does it really make sense to grow dirty things in order to get clean things? We have to be smarter than that.

Over and over again, it turns out that the dominant belief in the necessity of growth is under-justified. Those who call for continued growth at the expense of ecological stability are ready to risk everything – literally – for the sake of something we don't really even need.

We need new indicators of progress – but that's not enough

When Simon Kuznets introduced the GDP metric to the US Congress back in the 1930s, he was careful to warn that it should never be used as a normal measure of economic progress. Focusing on GDP would incentivise too much destruction. 'The welfare of a nation can scarcely be inferred from a measure of national income,' Kuznets said. 'Goals for more growth should specify more growth of what and for what.' A generation later, in 1968, the US politician Robert Kennedy conveyed this same message during a speech at the University of Kansas: 'GDP measures neither our wit nor our courage, neither our wisdom nor our learning, neither our compassion nor our devotion to our country … it measures everything, in short, except that which makes life worthwhile.'

And yet, nearly a century after Kuznets, and half a century after Kennedy, GDP remains the dominant measure of progress everywhere in the world. Kuznets opened Pandora's box, almost by accident, and no one has been able to close it since.

That's beginning to change, however. Growthism is starting to lose its ideological grip, even among some of the world's most prominent economists. In 2008, the French government established a high-level commission to define success in ways other than just GDP. In the same year, the OECD and the European Union launched their 'Beyond GDP' campaigns. As part of this effort, the Nobel Laureates Joseph Stiglitz and Amartya Sen published a report titled 'Mismeasuring our Lives: Why GDP Doesn't add Up'. In it, they took up Kuznets' plea and argued that over-reliance on GDP blinds us to what's happening to social and ecological health. The OECD launched a new metric off the back of this report, the Better Life Index, which incorporates welfare indicators like housing, jobs, education, health and happiness.

There is now a fast-growing list of alternative metrics, including the Index of Sustainable Economic Welfare and the Genuine Progress Indicator, both of which set out to correct GDP for social and ecological costs. And this new thinking is beginning to trickle into policy too. New Zealand's Prime Minister Jacinda Ardern captured headlines in 2019 with her promise to abandon GDP growth as an objective in favour of well-being. Nicola Sturgeon, the popular First Minister of Scotland, quickly followed suit, along with the Prime Minister of Iceland Katrín Jakobsdóttir. With each announcement, social media erupted with excitement and the stories went viral (and of course the fact that all three of these leaders are women did not escape notice). People are clearly ready for something different.

Suddenly, it's all the rage. And it's not just happening in rich countries. NGOs all over the world are now talking about the importance of 'well-being economies'. Countries like Bhutan, Costa Rica, Ecuador and Bolivia have all taken steps in this direction. And in 2013 the Chinese President Xi Jinping announced that GDP will no longer be used as the key metric of progress in China, reversing longstanding policy.

*

Adopting more holistic measures of progress is a crucial first step in the right direction. If politicians were to focus on maximising a measure like GPI, and if they were judged accordingly, they would be incentivised to improve social goods while curtailing ecological bads. It doesn't have to be GPI, though. It could be any of the alternative indicators that have been proposed. As soon as we shake ourselves free from the tyranny of GDP, we can have an open discussion about what we really value. This is the ultimate democratic act, and yet so far the ideological lockdown on growthism has effectively prevented us from doing so.

At the same time, we need to face up to the limits of this approach. While using better indicators might remove some of the political pressure for growth, it will not in and of itself arrest the rise of the juggernaut. Material and energy use doesn't rise only because politicians and economists pursue GDP growth. It rises because capitalism is organised around the imperative of constant expansion. We might choose to measure well-being, but if industrial activity keeps expanding behind the scenes, as it were, we'll still end up in ecological trouble. It's a bit like if you are trying to improve your physical health. If you switch from tracking your blood pressure to tracking your weekly pub quiz score, or the number of times you laugh each day, your life might well improve according to those other indicators, but your body might still be in trouble.

Here's the key point we need to grasp: GDP is not an *arbitrary* metric of economic performance. It's not as though it's some kind of mistake – an accounting error that just needs to be corrected. It was devised specifically in order to measure the welfare of capitalism. It externalises social and ecological costs because capitalism externalises social and ecological costs. It's naïve to imagine that if policymakers stop measuring GDP, capital will automatically cease its constant pursuit of ever-increasing returns, and our economies will become more sustainable. Those who call for a shift towards well-being as the sole solution tend to miss this point. If we want to release our society from the grip of the growth imperative, we have to be smarter than that.

Five

Pathways to a Post-Capitalist World

> We cannot save the world by playing by the rules. Because
> the rules have to be changed.
>
> Greta Thunberg

Once we understand that we can flourish without growth, our
horizons suddenly open up. It becomes possible to imagine a different kind of economy, and we're free to think more rationally
about how to respond to the climate emergency. It's a bit like
what happened during the Copernican Revolution. Early astronomers started from the assumption that the Earth sat at the
centre of the universe, but this caused endless amounts of
trouble: it meant that the movement of the other planets didn't
make any sense. It created mathematical problems that were
impossible to solve. When astronomers finally accepted that the
Earth and the other planets revolve around the Sun, suddenly
all the maths became *easier*. The same thing happens when we
take growth away from the centre of the economy. The ecological
crisis suddenly becomes much easier to solve.

Let's start with the most immediate challenge we face. The IPCC indicates that if we want to stay under 1.5°C (or even 2°C), without relying on speculative negative emissions technologies, then we need to scale down global energy use. Why? Because the less energy we use, the easier it is to achieve a rapid transition to renewables. Of course, low-income countries still need to increase their energy use in order to meet human needs. So it's high-income countries we need to focus on here; countries that consume vastly more than they require.

This is not just about individual behavior change, like turning off the lights when you leave a room. Sure, this kind of thing is important (and obviously we need to switch to LED bulbs, improve home insulation and so on), but ultimately we need to change how the economy works. Think of all the energy that's needed to extract and produce and transport all the stuff the economy churns out each year. It takes energy to pull raw materials out of the earth, and to power the factories that turn them into finished products. It takes energy to package those products and send them around the world on trucks and trains and aeroplanes, to build warehouses for storage and retail outlets for sales, and to process all the waste when they're binned. Capitalism is a giant energy-sucking machine.[1] In order to reduce energy use, we need to abandon aggregate growth as an objective, scale down less necessary forms of production, and organise the economy around supporting strong social outcomes.

This is what we mean by 'degrowth'. Degrowth is about reducing the material and energy throughput of the economy to bring it back into balance with the living world, while distributing income and resources more fairly, liberating people from needless work, and investing in the public goods that people need to thrive. It is the first step toward a more ecological civilisation. Of course, doing this may mean that GDP grows more slowly, or stops growing, or even declines. And if so, that's okay; because GDP isn't what matters. Under normal circumstances, this might cause a recession. But a recession is what happens when a growth-dependent economy stops growing: it's a disaster. Degrowth is completely different. It is about shifting to a different kind of economy altogether – an economy that doesn't *need* growth in the first place. An economy that's organised around human flourishing and ecological stability, rather than around the constant accumulation of capital.

The emergency brake

As we saw in Chapter 2, high-income nations consume on average 28 tons of material stuff per person per year. We need to bring that back down to sustainable levels.[2] What's brilliant about focusing on materials is that it has a range of powerful benefits. Slowing down material use means taking pressure off ecosystems. It means less deforestation, less habitat destruction, less biodiversity collapse. And it means our economy will use less energy, thus enabling us to achieve a faster transition to renewables. It also means we will need fewer solar panels and wind turbines and batteries than would otherwise be the case, which means less pressure on the places (mostly in the global South) where the materials for these things are extracted, and less pressure on the communities that live there.

In other words, degrowth – reducing material and energy use – is an ecologically coherent solution to a multi-faceted crisis. And the good news is that we can do this without any negative impact on human welfare. In fact, we can do it while *improving* people's lives.[3] How is this possible? The key is to remember that capitalism is a system that's organised around exchange-value, not around use-value. The majority of commodity production is geared toward accumulating profit rather than toward satisfying human needs. In fact, in a growth-oriented system, the goal is quite often to *avoid* satisfying human needs, and even to perpetuate need itself. Once we understand this, it becomes clear that there are huge chunks of the economy that are actively and intentionally wasteful, and which do not serve any recognisable human purpose.

Step 1. End planned obsolescence

Nowhere is this tendency clearer than when it comes to the practice of planned obsolescence. Companies desperate to increase sales seek to create products that are intended to break down and require replacement after a relatively short period of time. The practice was first developed in the 1920s, when lightbulb manufacturers, led by the US company General Electric, formed a cartel and plotted to shorten the lifespan of incandescent bulbs – from an average of about to 2,500 hours down to 1,000 or even less.[4] It worked like a charm. Sales shot up and profits soared. The idea quickly caught on in other industries, and today planned obsolescence is a widespread feature of capitalist production.

Take household appliances, for example – things like refrigerators, washing machines, dishwashers and microwaves. Manufacturers admit that the average lifespan of these products has dropped to less than seven years.[5] But when these products 'die' it's due not to system-wide failure, but rather to small electrical components that can easily be designed to last many years longer, at minimal cost. And yet to repair these parts is often prohibitively expensive, only marginally less than the cost of replacing the whole machine. Indeed, in many cases appliances are designed to lock mechanics out of the job altogether. People end up scrapping huge chunks of perfectly good metal and plastic every few years for no good reason at all.

The same is true of the technological devices we use every day. Anyone who has ever owned an Apple product knows this all too well. Apple's growth strategy seems to rely on a triple tactic: after a few years of use, devices become so slow as to be worthless; repairs are either impossible or prohibitively expensive; and advertising campaigns are designed to convince people that their products are obsolete anyhow. Apple is not the only one, of course. Tech companies sold a total of 13 billion smartphones between 2010 and 2019. Only about 3 billion of them are in use today.[6] That means 10 billion smartphones have been discarded over the past decade. Add desktops, laptops and tablets and we're talking about mountains of needless e-waste – most of it generated by planned obsolescence. Every year, 150 million discarded computers are shipped to countries like Nigeria, where they end up in sprawling open-air dumps that leak mercury, arsenic and other toxic substances into the land.[7]

It's not that the possibility for long-lasting, upgradable devices doesn't exist – it does – but its development is suppressed in favour of growth. Our biggest technology firms, which we celebrate as our greatest innovators, stifle the innovation we need because it runs *against the growth imperative*. And it's not just appliances and smartphones. It's everything. Nylon stockings that are designed to tear after a few wears, devices with new ports that render old dongles and chargers useless – everyone has stories about the absurdities of planned obsolescence. IKEA became a multi-billion-dollar empire in large part by inventing furniture that is effectively disposable. Whole swathes of Scandinavia's forests have been churned into cheap tables and shelving units that are designed for the dump.

There's a paradox here. We like to think of capitalism as a system that's built on rational efficiency, but in reality it is exactly the opposite. Planned obsolescence is a form of intentional *in*efficiency. The inefficiency is (bizarrely) rational in terms of maximising profits, but from the perspective of human need, and from the perspective of ecology, it is madness: madness in terms of the resources it wastes, and madness in terms of the needless energy it consumes. It is madness too in terms of human labour, when you consider the millions of hours that are poured into producing smartphones and washing machines and furniture simply to fill the void created, intentionally, by planned obsolescence. It's like shovelling ecosystems and human lives into a bottomless pit of demand. And the void will never be filled.

In a *genuinely* rational and efficient economy, companies like Apple would innovate to produce long-lasting, modular devices (like the Fairphone, for example), scale down their sales of new products, and maintain and upgrade existing stock wherever possible. But in a capitalist economy, this is not an option. Some might be tempted to blame individuals for buying too many smartphones or washing machines, but this misses the point. People become *victims* of this machine. Blaming individuals misdirects our attention away from the systemic causes.

How might we address these inefficiencies? One option is to introduce mandatory extended warranties on products. The technology already exists for appliances to last on average two to five times longer than they presently do, with lifespans up to thirty-five years, at little additional cost. With simple legislation, we could require manufacturers to guarantee their products for the duration of maximum feasible lifespans. If Apple was held to a 10-year guarantee, watch how quickly they would redesign their products to be resilient and upgradeable.

We could also introduce a 'right to repair', making it illegal for companies to produce things that can't be repaired by ordinary users, or by independent mechanics, with affordable replacement parts. Laws along these lines are already being considered in a number of European parliaments. Another option would be to switch to a lease model for large appliances and devices, requiring manufacturers to assume full responsibility for all repairs, with modular upgrades to improve efficiency whenever possible.

Measures like these would ensure that products (not just appliances and computers but furniture and houses and cars) would last many times longer than they presently do. And the effects would be significant. If washing machines and smartphones lasted four times longer, we would consume 75% fewer of them. That's a big reduction of material throughput, without any negative impact on people's lives. In fact, if anything it would *improve* quality of life, as people wouldn't have to deal with the frustration and expense of constantly replacing their equipment.

Step 2. Cut advertising

Planned obsolescence is only one of the strategies that growth-oriented firms use to speed up turnover. Advertising is another.

The advertising industry has seen wild changes over the past century. Up to the 1920s, consumption was a relatively perfunctory act: people just bought what they needed. Advertisements did little more than inform customers of the useful qualities of a product. But this system posed an obstacle to growth, because once people's needs were satisfied, purchases slowed down. Companies seeking a 'fix' – a way to surmount the limits of human need – found it in the new theories of advertising being developed at the time by Edward Bernays, the nephew of psychoanalyst Sigmund Freud. Bernays pointed out that you can provoke people to consume far beyond their needs simply by manipulating their psychology. You can seed anxiety in people's minds, and then present your product as a solution to that anxiety. Or you can sell things on the promise that they will provide social acceptance, or class distinction, or sexual prowess. This kind of advertising quickly became indispensable to American companies desperate to generate growing demand.

A survey conducted in the 1990s revealed that 90% of American CEOs believed it would be impossible to sell a new product without an advertising campaign; 85% admitted that advertising 'often' persuaded people to buy things they did not need; and 51% said that advertising persuaded people to buy things *they didn't actually want*.[8] These are extraordinary figures. They reveal that advertising amounts to manipulation on an industrial scale. And in the age of the internet, it has become more powerful and insidious than even Bernays himself could have dreamed. Browser cookies, social media profiles and big data allow firms to present us with ads tailored not just to our personalities – our specific anxieties and insecurities – but even to our likely emotional state at any given time. Firms like Google and Facebook are worth more than companies like BP and Exxon, purely on the promise of advertising. We think of these companies as innovators, but the majority of their innovations appear to be focused on developing ever more sophisticated tools to get people to buy things.

It's a kind of psychological warfare. Just as the oil industry has turned to more aggressive ways of extracting reserves that are increasingly difficult to reach, so too advertisers are turning to more aggressive ways of getting at the last remaining milliseconds of our attention. They are fracking, as it were, for our minds. We are exposed to thousands of ads every day, and with every year that ticks by the ads become more insidious. It's an assault on our consciousness – the colonisation not only of our public spaces but also of our minds. And it works. Research reveals that advertising expenditures have a direct and highly significant impact on material consumption.[9] The higher the spend, the higher the consumption. And right now the global advertising spend is rising fast: from $400 billion in 2010, to $560 billion in 2019, making it one of the biggest industries in the world.[10]

Sometimes advertising unites with planned obsolescence in a toxic cocktail. Take the fashion industry, for example. Clothing retailers desperate to increase sales in an over-saturated market have turned to designing clothes that are *meant to be discarded* – cheap, flimsy garments that last only for a few wears, and are intended to 'go out of style' within months. Ads are deployed to convince people that the clothes they own are dull, outdated and inadequate (a tactic sometimes referred to as 'perceived obsolescence'). The average American today purchases five times as many garments each year as they did in 1980. In the UK, textile purchases surged by 37% in the four years from 2001 to 2005, as 'fast-fashion' techniques exploded into the mainstream.[11] The industry's material use has skyrocketed to over 100 million tons per year, and energy, water and land use have soared right along with it.

If we take the American data as a standard, we can assume that regulations against fast fashion alone could in theory cut textile throughput by up to 80%, without compromising people's access to the clothes they need.

There are many ways to curb the power of advertising. We can introduce quotas to reduce total ad expenditure. We can legislate against the use of psychologically manipulative techniques. And we can liberate public spaces from ads – both offline and online – where people don't have a choice about what they see. São Paulo, a city of 20 million people, has already done this in key parts of the city. Paris has made moves in this direction too, reducing outdoor ads and even banning them outright in the vicinity of schools. The results? Happier people: people who feel more secure about themselves and more content with their lives. Cutting ads has a direct positive impact on people's well-being.[12] In addition to slowing down needless consumption, these measures would also free our minds – so we can follow our thoughts, our imaginations, our creativity without being constantly interrupted. And we can fill those spaces instead with art and poetry, or with messages that build community and affirm intrinsic values.

Some economists worry that limiting advertising would undermine market efficiency. Ads help people make rational decisions about what to buy, they say. But this claim doesn't hold water. In reality, most advertising does exactly the opposite: it is designed to manipulate people into making *irrational* decisions.[13] And let's face it: in the age of the internet, people don't actually need ads in order to find and evaluate products. A simple search is enough to do the trick. The internet has rendered advertising obsolete (ironically, for a place that has become filled with ads), and we should embrace this fact.

Step 3. Shift from ownership to usership

There is another inefficiency that's built into capitalism. A lot of the stuff we consume is necessary but rarely used. Pieces of equipment like lawnmowers and power tools are used perhaps once a month, for maybe an hour or two at most, and for the rest of the year lie idle. Manufacturers want everyone to own a garage full of things that can otherwise quite easily be shared, but a more rational approach would be to establish neighbourhood workshops where equipment can be stored and used on an as-need basis. Some communities are already doing this, maintaining shared equipment with a neighbourhood fund. Projects like these can be scaled up by city governments, and enabled by apps for easy access. Shifting from ownership to 'usership' can have a big impact on material throughput. Sharing a single piece of equipment among ten households means cutting demand for that product by a factor of ten, while saving people time and money in the process.

This is particularly true of cars. We know we need to switch to electric cars, but ultimately we also need to dramatically scale down the total number of cars. The most powerful intervention by far is to invest in affordable (or even free) public transportation, which is more efficient in terms of the materials and energy required to move people around. This is vital for any plan to get off fossil fuels. Bicycles are even better, as many European cities are learning (as I write this, Milan is handing over 35 kilometres of streets to cyclists, in a bid to keep pollution low after their coronavirus lockdown). And for journeys that can't be made with either, we can develop publicly owned, app-based platforms for sharing cars between us – without the rentier intermediation that has made platforms like Uber and Airbnb so problematic.

Step 4. End food waste

Here's a fact that never ceases to amaze me: up to 50% of all the food that's produced in the world – equivalent to 2 billion tonnes – ends up wasted each year.[14] This happens across the supply chain. In high-income nations it's due to farms that discard vegetables that aren't cosmetically perfect, and supermarkets that use unnecessarily strict sell-by dates, aggressive advertising, bulk discounts and buy-one-get-one-free schemes. Households end up tossing away 30-50% of the food they purchase. In low-income nations it's due to poor transportation and storage infrastructure, which means food ends up rotting before it makes it to market.

Food waste represents an extraordinary ecological cost, in terms of energy, land, water and emissions. But it also represents a big opportunity. Ending food waste could in theory cut the scale of the agriculture industry in half, without any loss of access to the food we presently need. That would allow us to cut global emissions by up to 13%, while regenerating up to 2.4 billion hectares of land for wildlife habitat and carbon sequestration.[15]

When it comes to degrowth, this is low-hanging fruit. Some countries are already taking steps in this direction. France and Italy have both recently passed laws preventing supermarkets from wasting food (they have to donate unsold food to charities instead). South Korea has banned food waste from landfills altogether, and requires households and restaurants to use special composting receptacles that charge fees by weight.

Step 5. Scale down ecologically destructive industries

On top of targeting intentional inefficiencies and waste, we also need to talk about scaling down specific industries that are ecologically destructive and socially less necessary. The fossil fuel industry is the most obvious example, but we can extend this logic to others.

Take the beef industry, for instance. Nearly 60% of global agricultural land is used for beef – either directly for cattle pasture or indirectly for growing feed.[16] It's one of the most resource-inefficient foods on the planet, in terms of the land and energy it requires per calorie or nutrient. And the pressure to find land for pasture and feed is the single greatest driver of deforestation. As I write this, large parts of the Amazon rainforest are literally being burned down for the sake of beef. Yet, far from being essential to human diets, beef accounts for only 2% of the calories humans consume. In most cases the industry could be radically scaled down without any loss to human welfare.[17]

The gains would be astonishing. Switching from beef to non-ruminant meats or plant proteins like beans and pulses could liberate almost 11 million square miles of land – the size of the United States, Canada and China *combined*.[18] This simple shift would allow us to return vast swathes of the planet to forest and wildlife habitat, creating new carbon sinks and cutting net emissions by up to 8 gigatons of carbon dioxide per year, according to the IPCC. That's around 20% of current annual emissions. Scientists say that degrowing the beef industry is among the most transformative policies we could implement, and is essential to avoiding dangerous climate change.[19] A first step would be to end the subsidies high-income countries give to beef farmers. Researchers are also testing proposals for a tax on red meat, which they find would not only curtail emissions but deliver a wide range of public health benefits, while driving medical costs down.[20]

The beef industry is just one example. There are many others we could consider. We could scale down the arms industry and the private jet industry. We could scale down the production of single-use plastics, disposable coffee cups, SUVs and McMansions (in the United States, house sizes have doubled since the 1970s[21]). Instead of building new stadiums for the Olympics and the World Cup every few years we could reuse existing infrastructure. We know that to reach our climate goals we will need to scale down the commercial airline industry, starting with policies like a frequent flyer levy, ending routes that can be served by train, and getting rid of first-class and business-class cabins, which have the highest CO_2 per passenger mile. And we must shift from an economy based on energy-intensive long-distance supply chains to one where production happens closer to home.

We need to have an open, democratic conversation about this. Rather than assuming that all sectors must grow, for ever, regardless of whether or not we actually need them, let's talk about what we want our economy to deliver. What industries are already big enough and shouldn't grow any larger? What industries could be usefully scaled down? What industries do we still need to expand? We have never asked these questions. During the coronavirus pandemic in 2020, we all learned the difference between 'essential' industries and superfluous ones; it quickly became apparent which industries are organised around use-value, and which ones are mostly about exchange-value. We can build on those lessons.

*

This is not meant to be an exhaustive list. My point here is to illustrate that we can accomplish significant reductions in material throughput without any negative impact on human welfare. And here's the powerful part. This approach would not only reduce the *flows* of material goods, it would also reduce the *stocks* that support those flows. Half of all the materials that we extract each year go to building up and maintaining material stocks: things like factories and machines and transport infrastructure.[22] If we consume half as many products, we also need half as many factories and machines to produce them, half as many aeroplanes and trucks and ships to transport them, half as many warehouses and retail outlets to distribute them, half as many garbage trucks and waste disposal plants to process them when they're binned, and half as much energy to produce and maintain and operate all of that infrastructure. The efficiencies begin to multiply.

Ultimately, governments need to set concrete targets for reducing material and energy use. As we saw in Chapter 3, taxes alone won't be enough. Ecological economists insist that the only way to do it is with a hard limit: cap resource and energy use at existing levels and ratchet them down each year until you get back within planetary boundaries.[23] There's nothing particularly radical about this; after all, we place all sorts of limits on capital's exploitation of people, including minimum wage laws, child labour laws and the weekend. So too we need to place limits on capital's exploitation of nature.

The key is that this has to be done in a just and equitable way, to ensure that everyone has access to the resources and livelihoods they need to flourish, and so small businesses don't get squeezed out by bigger players. This can be done with a cap, fee and dividend system: charge industries a progressively rising fee for resource and energy use, and distribute the yields as an equal dividend to all citizens. The Yellow Vests movement that erupted in France in 2018 rightly rejected the government's attempts to balance environmental goals on the backs of the working class and poor. Injustice cannot solve a problem that has been caused by injustice in the first place. We need to take the opposite approach.

But what about jobs?

Now, here's where things get tricky. The policies I've suggested above are likely to reduce total industrial production. This might be OK from the perspective of human needs (none of us would be worse off if our smartphones lasted twice as long), but it does leave us with a difficult question. As products last longer, as we shift to sharing things, and as we slash food waste and scale down fast fashion, employment in these industries will decline and jobs will disappear across the supply chains. In other words, as our economy becomes more rational and efficient, it will require less labour.

From one perspective, this is fantastic news. It means that fewer lives will be wasted in needless jobs, producing and selling things that society doesn't actually require. It means liberating people to spend their time and energy on other things. But from the perspective of the individual workers who will be laid off from these jobs, it is a disaster. And governments will find themselves struggling to cope with unemployment.

This might seem like an impossible bind; and indeed it's one reason why politicians consider degrowth to be so unthinkable. But there's a way out. As we shed unnecessary jobs we can shorten the working week, going from forty-seven hours (the average in the United States) down to thirty or perhaps even twenty hours, sharing necessary labour more evenly among the working population and maintaining full employment. This approach would allow everyone to benefit from the time that's liberated by degrowth. And retraining programmes can be deployed to ensure that people are able to transition easily from sunset industries to other kinds of work, so no one gets left out. We can facilitate this process by introducing a public job guarantee (a policy that happens to be resoundingly popular[24]), so that any-one who wants to work can get a job doing socially useful things that communities actually need, like care, essential services, building renewable energy infrastructure, growing local food, and regenerating degraded ecosystems – paid at a living wage.[25] Indeed, a job guarantee is one of the single most powerful envir-onmental policies a government could implement, because it enables us to have an open conversation about scaling down destructive industries without worrying about the spectre of unemployment.

The exciting part is that reducing working hours has a substan-tial positive impact on people's well-being. This effect has been demonstrated over and over again, and the results are striking. Studies in the US have found that people who work shorter hours are happier than those who work longer hours, even when controlling for income.[26] When France downshifted to the thirty-five-hour week, workers reported that their quality of life improved.[27] An experiment in Sweden showed that employees who reduced their working time to thirty hours reported improved life satisfaction and better health outcomes.[28] Data also shows that shorter hours leave people feeling more satisfied with their jobs, boosting morale and happiness.[29] And – perhaps best of all – shorter hours are associated with greater gender equality, both in the workplace and at home.[30]

Some critics worry that if you give people more time off they'll spend it on energy-intensive leisure activities, like taking long-haul flights for holidays. But the evidence shows exactly the opposite. It is those with *less* leisure time who tend to consume more inten-sively: they rely on high-speed travel, meal deliveries, impulsive purchases, retail therapy, and so on. A study of French households found that longer working hours are directly associated with higher consumption of environmentally intensive goods, even when correcting for income.[31] By contrast, when people are given time off they tend to gravitate towards lower-impact activities: exercise, volunteering, learning, and socialising with friends and family.[32]

These effects play out across whole countries. For instance, researchers have found that if the United States were to reduce its working hours to the levels of Western Europe, its energy consumption would decline by a staggering 20%. Shortening the working week is one of the most immediately impactful climate policies available to us.[33]

But perhaps the most important part about shortening the working week is that it frees people to spend more time *caring* – be it nursing a sick relative, playing with children, or helping restore a woodland. This essential reproductive work (most of which is normally done by women) is totally devalued under capitalism; it is externalised, unpaid, invisible and unrepresented in GDP figures. Degrowth will free us to reallocate labour to what really matters – to things that have real use-value. Care contributes directly to social and ecological well-being, and participating in caring activities has been shown to be more powerful than material consumption when it comes to improving people's sense of happiness and meaning, vastly outstripping the dopamine hit we might get from a shopping binge.

The benefits of a shorter working week keep multiplying. One group of scientists summed up the evidence like this: 'Overall, the existing research suggests that working time reduction potentially offers a triple dividend to society: reduced unemployment, increased quality of life, and reduced environmental pressures.'[34] Transitioning to a shorter working week is key to building a humane, ecological economy.

*

There's nothing new about this idea. In fact, it's not even particularly radical. In 1930, the British economist John Maynard Keynes wrote an essay titled 'Economic Possibilities for Our Grandchildren'. He predicted that by the year 2030 technological innovation and improvements in labour productivity would free people to work only fifteen hours a week. Keynes turned out to be correct about productivity gains, but his prophecy about working hours never came true. Why not? Because gains in labour productivity have been appropriated by capital. Instead of shortening the working week and raising wages, companies have pocketed the extra profits and required employees to keep working just as much as before. In other words, productivity gains have been used not to liberate humans from work but rather to fuel constant growth.

In this sense, capitalism betrays the very Enlightenment values it claims to advance. We normally think of capitalism as organised around the principles of freedom and human liberation – that's the ideology it sells us. And yet while capitalism has produced the technological capacity to provide for everyone's needs many times over, and to liberate people from unnecessary labour, it deploys that technology instead to create new 'needs' and to endlessly expand the treadmill of production and consumption. The promise of true freedom is perpetually deferred.[35]

Reduce inequality

People often wonder whether there will be enough income in a degrowth scenario for everyone to meet their needs. The answer is yes, by definition. National income is the obverse of the prices of all the goods that the economy produces. As long as we are producing what people need, there will always be exactly enough income to buy those things. What matters is *distribution*. Ensuring a fair distribution of income is everything.

Some of this will be accomplished automatically with the shorter working week and the job guarantee. These measures would dramatically improve the bargaining power of labour, and increase workers' share of the national income. But we can also add another layer of protection, by introducing a living wage policy that's pegged to the week or month, rather than to the hour. In a degrowth scenario, this means shifting income from capital back to labour, reversing the appropriation of productivity gains that has happened since Keynes penned his essay in 1930. A shorter working week would be funded, in other words, by reducing inequality.

There's plenty of room for this. In the UK, labour's share of national income has declined from 75% in the 1970s down to only 65% today. In the United States it's down to 60%. Hourly wages at the bottom could be raised quite a lot by reversing these losses. There's plenty of room for this within companies too. CEO compensation has grown to dizzying heights in recent decades, with some executives capturing as much as $100 million per year. And the gap between CEO salaries and the wages of average workers has exploded. In 1965, CEOs earned about twenty times more than the average worker. Today they earn on average 300 times more.[36] And in some companies the gap is even more extreme. In 2017, Steve Easterbrook, the CEO of McDonald's, earned $21.7

million while the median full-time McDonald's worker earned $7,017. That's a ratio of 3,100 to one. In other words, the average McDonald's employee would have to work 3,100 years – every day from the advent of ancient Greece until now – to earn what Steve Easterbrook received in his annual pay cheque.[37]

One approach would be to introduce a cap on wage ratios: a 'maximum wage' policy. Sam Pizzigati, an associate fellow at the Institute for Policy Studies, argues that we should cap the after-tax wage ratio at 10 to 1.[38] CEOs would immediately seek to raise wages as high as they can reasonably go. It's an elegant solution, and it's not unheard of. Mondragon, a huge workers' co-operative in Spain, has rules stating that executive salaries cannot be more than six times higher than the lowest-paid employee in the same enterprise. Better yet, we could do it on a national scale, by saying that incomes higher than a given multiple of the national minimum wage would face a 100% tax. Imagine how quickly the income distribution would change.

But it's not just income inequality that's a problem – it's *wealth* inequality too. In the United States, for instance, the richest 1% have nearly 40% of the nation's wealth. The bottom 50% have almost nothing: only 0.4%.[39] On a global level the disparities are even worse: the richest 1% have nearly 50% of the world's wealth. The problem with this kind of inequality is that the rich become extractive rentiers. As they accumulate money and property far beyond what they could ever use, they rent it out (be it residential or commercial properties, patent licences, loans, whatever). And because they have a monopoly on these things, everyone else is forced to pay them rents and debts. This is called 'passive income', because it accrues automatically to people who hold capital without any labour on their part. But from the perspective of everyone else it is anything but passive: people have to scramble to work and earn above and beyond what they would otherwise need, simply in order to pay rents and debts to the rich. It is like modern-day serf-dom. And just like serfdom, it has serious implications for our living world. Serfdom was an ecological disaster because lords forced peasants to extract more from the land than they otherwise needed – all in order to pay tribute. This led to a progressive degra-dation of forests and soils. So it goes today: we are made to plunder the Earth simply to pay tribute to millionaires and billionaires.

One way to solve this problem is with a wealth tax (or a solidarity tax, perhaps). The economists Emmanuel Saez and Gabriel Zucman have proposed a 10% annual marginal tax on wealth holdings over $1 billion. This would push the richest to sell some of their assets, thus distributing wealth more fairly. But in an era of ecological crisis, we must be more ambitious than this. After all, nobody 'deserves' this kind of wealth. It's not earned, it's extracted: from underpaid workers, from cheap nature, from rent-seeking, from political capture and so on. Extreme wealth has a corrosive effect on our society, on our political system, and on the living world. We should have a democratic conversation about this: at what point does hoarding become destructive and unacceptable? $100 million? $10 million? $5 million?

As we saw in the previous chapter, reducing inequality is a powerful way to reduce ecological pressure. It cuts high-impact luxury consumption by the rich, and reduces competitive con-sumption across the rest of society. But it also removes pressures for unnecessary growth. The policies I've proposed here would lead to a disaccumulation of capital. This would cut rent-seeking behaviour, and the rich would lose their power to force us to extract and produce more than we need. The economy would shift away from unnecessary exchange-value and more towards use-value. It would also reduce political capture and improve the quality of democracy; and democracy, as we will see later, has intrinsic ecological value.

Decommodify public goods and expand the commons

As we scale down excess industrial production we can mitigate impact on livelihoods by distributing labour, income and wealth more fairly. But there's another crucial point to add. Remember, when it comes to human welfare, it's not income itself that matters; it's the welfare purchasing power of income that counts.

Let's take an example that's close to my own experience: housing in London. House prices are astronomically high, to the point where a normal two-bedroom flat may cost £2,000 a month to rent, or £600,000 to buy. These prices bear no relationship to the cost of the land, materials and labour involved in building a house. They're a consequence of policy decisions, such as the privatisation of public housing since 1980, and the low interest rates and quantitative easing that have pumped up asset prices since 2008. Meanwhile, wages in London have not kept pace – not even close. To cover the gap, ordinary Londoners have had to either work longer hours or take out loans (which represent a claim on their future labour), just to access a basic social good they used to be able to get for a fraction of the cost. In other words, as house prices have soared, the welfare purchasing power of Londoners' incomes has declined.

Now, imagine we drive rents down with permanent rent controls (a policy that 74% of British people happen to support[40]). Prices would still be outrageously high, but suddenly Londoners would be able to work and earn less than they presently do *without any loss to their quality of life*. Indeed, they would *gain* in terms of extra time to spend with family, hanging out with friends, and doing things they love.

We could do the same thing with other goods that are essential to people's well-being. Healthcare and education are obvious ones. But why not the internet? Why not public transport? Why not basic quotas of energy and water? Researchers at the University of London have demonstrated that a full range of what they call Universal Basic Services could be publicly funded (with progressive taxation on wealth, land, carbon, etc.) at costs much lower than we presently spend, while guaranteeing everyone access to a decent, dignified life.[41] On top of this, we could invest in public libraries, parks and sports grounds. Facilities like these become particularly important as we shorten the working week, so that people can spend their time in ways that enrich well-being with little environmental impact.[42]

Decommodifying basic goods and expanding the commons allows us to improve the welfare purchasing power of incomes, so people can access the things they need to live well without needing ever-higher incomes in order to do so. This approach reverses the Lauderdale Paradox we explored in Chapter 1. Capitalists enclose commons ('public wealth') in order to generate growth ('private riches'), forcing people to work more simply to pay for access to resources they once enjoyed for free. As we create a post-growth economy we can flip this equation around: we can choose to restore commons, or create new commons, in order to render ever-rising incomes unnecessary. The commons become an antidote to the growth imperative.

A theory of radical abundance

This brings us to the real heart of a post-capitalist economy. Ending planned obsolescence, capping resource use, shortening the working week, reducing inequality and expanding public goods – these are all essential steps to reducing energy demand and enabling a faster transition to renewables. But they are also more than that. They fundamentally alter the deep logic of capitalism.

In Chapter 1 we saw how the rise of capitalism depended on the creation of artificial scarcity. From the enclosure movement to colonisation, scarcity had to be *created* in order to get people to submit to low-wage labour, to pressure them to engage in competitive productivity, and to recruit them as mass consumers. Artificial scarcity served as the engine of capital accumulation. This same logic operates today. It's all around us. Take the labour market, for example. People feel the force of scarcity in the constant threat of unemployment. Workers must become ever more disciplined and productive at work or else lose their jobs to someone who will be more productive still – usually someone poorer or more desperate. But as productivity rises, workers get laid off – and governments have to scramble for ways to grow the economy in order to create new jobs. Workers themselves join in the chorus calling for growth, and push to elect politicians who promise it. But it doesn't have to be this way. We *could* deliver productivity gains back to workers in the form of higher wages and shorter hours. The constant threat of unemployment is due to an *artificial* scarcity of jobs.

We see the same thing happening when it comes to the distribution of income. The vast majority of new income from growth gets siphoned straight into the pockets of the rich, while wages stagnate and poverty persists. Politicians and economists call for more growth in order to solve these problems, and everyone who is moved by the tragedy of poverty lines up behind them. But it never works as they promise it will, because the yields of growth trickle down so slowly, if at all. Inequality perpetuates an artificial scarcity of income.

This plays out in the realm of consumption too. Inequality stimulates a sense of inadequacy. It makes people feel that they need to work longer hours to earn more income to buy unnecessary stuff, just so they can have a bit of dignity.[43] In this sense, inequality creates an artificial scarcity of well-being. In fact, this effect is quite often wielded as an intentional strategy by economists and politicians. The British Prime Minister Boris Johnson once stated that 'inequality is essential for the spirit of envy' that keeps capitalism chugging along.

Planned obsolescence is another strategy of artificial scarcity. Retailers seek to create new needs by making products artificially short-lived, to keep the juggernaut of consumption from grinding to a halt. The same goes for advertising, which stimulates an artificial sense of lack; a sense that something is literally missing. Ads create the impression that we are not beautiful enough, or masculine enough, or stylish enough.

And then there's the artificial scarcity of time. The structural compulsion to work unnecessarily long hours leaves people with so little time that they have no choice but to pay firms to do things they would otherwise be able to do themselves: cook their food, clean their homes, play with their children, care for their elderly parents. Meanwhile, the stress of overwork creates needs for anti-depressants, sleep aids, alcohol, dieticians, marital counselling, expensive holidays, and other products people would otherwise be less likely to require. To pay for these things, people need to work yet more to increase their incomes, driving a vicious cycle of unnecessary production and consumption.

We see artificial scarcity being imposed on our public goods too. Since the 1980s endless waves of privatisation have been unleashed all over the world, of education, healthcare, transport, libraries, parks, swimming pools, water, housing, even social security. Social goods everywhere are under attack for the sake of growth. The idea is that by making public goods scarce, people will have no choice but to purchase private alternatives. And in order to pay, they will have to work more, producing additional goods and services that must find a market, and thereby creating new pressures for additional consumption elsewhere in the system.

This logic reaches its height in the politics of austerity, which was rolled out across Europe in the wake of the 2008 financial crisis. Austerity (which is literally a synonym for scarcity) is a desperate attempt to restart the engines of growth by slashing public investment in social goods and welfare protections – everything from elderly heating allowances to unemployment benefits to public sector wages – chopping away at what remains of the commons so that people deemed too 'comfortable' or 'lazy' are placed once again under threat of hunger, and forced to increase their productivity if they want to survive. This logic is overt, just as it was in the eighteenth and nineteenth centuries. During the government of British Prime Minister David Cameron, welfare cuts were conducted explicitly in order to get 'shirkers' to work harder and to be more productive ('workfare', they called it).

Over and over again, it becomes clear that scarcity is *created*, intentionally, for the sake of growth. Just as during the enclosures in the 1500s, scarcity and growth emerge as two sides of the same coin.

*

This exposes a remarkable illusion at the heart of capitalism. We normally think of capitalism as a system that generates so *much* (just consider the extraordinary cornucopia of stuff that's displayed on television and in shopfronts). But in reality it is a system that is organised around the constant production of scarcity. Capitalism transforms even the most spectacular gains in productivity and income not into abundance and human freedom, but into new forms of artificial scarcity. It must, or else it risks shutting down the engine of accumulation itself. In a growth-oriented system, the objective is not to satisfy human needs, but to *avoid* satisfying human needs. It is irrational and ecologically violent.

Once we grasp how this works, solutions rush into view. If scarcity is created for the sake of growth, then by reversing artificial scarcities we can render growth unnecessary. By decommodifying public goods, expanding the commons, shortening the working week and reducing inequality, we can enable people to access the goods that they need to live well without requiring additional growth in order to do so. People would be able to work less without any loss to their well-being, thus producing less unnecessary stuff and generating less pressure for unnecessary consumption elsewhere. And with our extra free time we would no longer have to engage in the patterns of consumption that are necessitated by time scarcity.[44]

Liberated from the pressures of artificial scarcity, and with basic needs met, the compulsion for people to compete for ever-increasing productivity would wither away. The economy would produce less as a result, yes – but it would also *need* less. It would be smaller and yet nonetheless much more abundant. In such an economy private riches (or GDP) may shrink, reducing the incomes of corporations and the elite, but public wealth would increase, improving the lives of everyone else. Exchange-value might go down, but use-value will go up. Suddenly a new paradox emerges: *abundance* is revealed to be the antidote to growth. In fact, it neutralises the growth imperative itself, enabling us to slow down the juggernaut and release the living world from its grip. As Giorgos Kallis has pointed out, 'capitalism cannot operate under conditions of abundance'.[45]

Some critics have claimed that degrowth is nothing more than a new version of austerity. But in fact exactly the opposite is true. Austerity calls for scarcity in order to generate more growth. Degrowth calls for abundance *in order to render growth unnecessary*. If we are to avert climate breakdown, the environmentalism of the twenty-first century must articulate a new demand: a demand for radical abundance.

The Law of Jubilee

Reversing artificial scarcity is a powerful step towards liberating us from the tyranny of growth. But there are also other pressures we have to deal with – other growth imperatives to neutralise.

Perhaps the most powerful of these is debt. If you're a student who wants to go to university, or a family that wants to buy a home, you might have to take out loans to do so. And the thing about loans is that they come with interest, and interest is a compound function that makes debt grow exponentially. When you owe debts to private creditors you can't just be satisfied with earning back as much as you borrowed; you have to find ways to grow your earnings fast enough to pay off growing debt. You may end up having to pay off your original loan many times over – perhaps even for the rest of your life. If you don't, then debt piles up and eventually triggers a financial crisis. Either you grow or you collapse.

Compound interest creates a kind of artificial scarcity. And it has a direct ecological impact. Countries loaded with external debts are under heavy pressure to deregulate logging and mining and other extractive industries, plundering ecosystems in order to meet their debt obligations (this is not true for deficits that governments owe to their own central banks, however; unlike external debts, these don't have to be repaid). The same is true of households. Researchers have found that households with high-interest mortgages work longer hours than they would otherwise need to simply in order to stay afloat.[46] As the anthropologist David Graeber has observed, the financial imperatives of debt 'reduce us all, despite ourselves, to the equivalent of pillagers, eyeing the world simply for what can be turned into money'.[47]

Fortunately, there's a way to relieve this pressure. We can just cancel some of the debt. In an era of ecological breakdown, debt cancellation becomes a vital step towards a more sustainable economy. This may sound radical, but there's plenty of precedent for it. Ancient Near-Eastern societies regularly declared non-commercial debts void, clearing the books and liberating people from bondage to creditors. This principle was institutionalised in the Hebrew Law of Jubilee, which decreed that debts should be automatically cancelled every seventh year.[48] Indeed, debt cancellation became core to the Hebrew concept of redemption itself.

There are dozens of proposals for how we might do this in today's economy. The US presidential candidate Bernie Sanders laid out a clear plan for cancelling student debts, which in 2020 stood at a staggering $1.6 trillion. Academics at King's College London have published a plan for how governments could write off not just student debts but also other unjust debts: mortgage debts created by housing speculation and quantitative easing, old debts whose lenders have been bailed out by governments, and unpayable debts that are devalued on secondary markets.[49] We know it's possible. In the wake of the coronavirus disaster in 2020, governments in a number of countries suddenly found the ability to make debts disappear.

We can do the same thing with the external debts held by global South countries, which have been rising at an alarming rate. Big chunks of that debt are holdovers from the 1980s, when the US Federal Reserve raised interest rates so high as to put whole countries into permanent bondage to Wall Street.[50] Then there are debts that were sold by corrupt lenders, and debts accumulated by old dictators with no democratic mandate who have long since been deposed. Researchers with the Jubilee Debt Campaign have proposed clear mechanisms for cancelling unjust debts like these, which would liberate poor countries from the pressure to plunder their own resources and exploit their citizens in the constant hunt for growth. Indeed, this is an important first step towards the reparations that rich countries owe for the climate debts they hold with respect to the rest of the world.

Big creditors would lose out, of course, but we might decide that this is OK – a loss we're willing to have them bear in order for us to build a fairer and more ecological society. We can cancel debts in such a way that nobody gets hurt.[51] Nobody dies. Compound interest is just a fiction, after all. And the nice thing about fictions is that we can change them. Perhaps no one has put this more eloquently than David Graeber:

> [Debt cancellation] would be salutary not just because it would relieve so much genuine human suffering, but also because it would be our way of reminding ourselves that money is not ineffable, that paying one's debts is not the essence of morality, that all these things are human arrangements and that if democracy is going to mean anything, it is the ability to all agree to arrange things in a different way.[52]

New money for a new economy

But debt cancellation is just a one-off fix; it doesn't really get to the root of the problem. There's a deeper issue we need to address.

The main reason our economy is so loaded with debt is because it runs on a money system that *is itself debt*. When you walk into a bank to take out a loan, you might assume that the bank is lending you money it has in its reserve, collected from other people's deposits and stored in a basement vault somewhere. But that's not how it works. Banks are only required to hold reserves worth about 10% of the money they lend out, or even less. This is known as 'fractional reserve banking'. In other words, banks lend out about ten times more money than they actually have. So where does that extra money come from, if it doesn't actually exist? Banks create it out of thin air when they credit your account. They literally *loan it into existence*.

More than 90% of the money that's presently circulating in our economy is created in this manner. In other words, almost every single dollar that passes through our hands represents somebody's debt. And this debt has to be paid back *with interest* – with more work, more extraction and more production. This is extraordinary, when you think about it. It means that banks effectively sell a product (money) that they produce out of nothing, for free, and then require people to go out into the real world and extract and produce real value to pay for it. It is so outlandish as to offend common sense. People have a difficult time believing it could possibly be true. As Henry Ford put it in the 1930s: 'It is perhaps well enough that the people of the nation do not know or understand our banking and monetary system, for if they did I believe there would be a revolution before tomorrow morning.'

Now, here's the problem. Banks create the principal for all the loans they give, but they don't create the money needed to pay the interest. There is always a deficit, always a scarcity. This scarcity creates intense competition, forcing everyone to scramble to find ways to get the money to pay back their debts, including by taking out yet more debt.

If you've ever watched a game of musical chairs, you have an idea of how this plays out. Each round of the game ramps up the scarcity of chairs, and players have to fight each other to get to one of the few that are left. It's chaos. Now imagine we up the stakes. Instead of just getting knocked out of the game, you lose your home, your kids go hungry, and you can't pay for medicines. Think about what such a game would look like – the desperate measures people would take to get to a chair – and you have a rough picture of how our economy works.[53] Casual observers of capitalist societies might conclude – as many economists have done – that vicious competition, maximisation and self-interested behaviour are hard-wired into human nature. But is it really human nature that makes us behave this way? Or is it just the rules of the game?

Over the past decade ecological economists have concluded that a money system based on compound interest is incompatible with sustaining life on a delicately balanced living planet. As for what to do about it, there are several ideas floating around. One group argues that all we need to do is switch from the existing compound interest system, where debt grows exponentially, to a simple interest system, where it grows linearly – adding the same increment each year. Over time this would put a huge dent in total debt levels, bring our money system back in line with ecology, and allow us to shift to a post-growth economy without causing a financial crisis.[54]

A second group argues that we need to go further, and abolish debt-based currency altogether. Instead of letting commercial banks create credit money, we could have the state create it – free of debt – and then *spend* it into the economy instead of *lending* it into the economy. The responsibility for money creation could be placed with an independent agency that is democratic, accountable and transparent, with a mandate to balance human well-being with ecological stability. Banks would still be able to lend money, of course, but they would have to back it with 100% reserves, dollar for dollar.[55]

This is not a fringe idea. It was first proposed by economists at the University of Chicago in the 1930s, as a solution to the debt crisis of the Great Depression. It made headlines again in 2012 when it was promoted by some progressive IMF economists as a way of reducing debt and making the global economy more stable. In the United Kingdom, a campaigning group called Positive Money has built a movement around the idea, and now it's being picked up as another possible step towards a more ecological economy. What's powerful about this approach isn't just that it reduces debt, but that a public money system would enable us to fund things like universal healthcare, a job guarantee, ecological regeneration and energy transition *directly*, without having to chase GDP growth in order to generate revenues.[56]

A post-capitalist imaginary

When people talk about 'overthrowing' or 'abolishing' capitalism, it can leave us with a real sense of unease about what will come afterwards. It's easy to feel angry about our economic system, especially as we watch our planet die, but those who call for revolution all too rarely define what the new society might look like. It makes the future seem scary and unpredictable – who knows what nightmares might fill the void?

But when we focus on how to release our system from the growth imperative, we begin to get a sense of what a post-capitalist economy might look like. And it's not scary at all. This is not the command-and-control fiasco of the Soviet Union, or some back-to-the-caves, hair-shirted disaster of voluntary impoverishment. On the contrary, it's an economy that feels in key ways *familiar*, in the sense that it resembles the economy as we normally describe it to ourselves (in other words, perhaps as we wish it to be): an economy where people produce and sell useful goods and services; an economy where people make rational, informed decisions about what to buy; an economy where people get compensated fairly for their labour; an economy that satisfies human needs while minimising waste; an economy that circulates money to those who need it; an economy where innovation makes better, longer-lasting products, reduces ecological pressure, frees up labour time and improves human welfare; an economy that responds to – rather than ignores – the health of the ecology on which it depends.

And yet inasmuch as it is familiar in these ways, the new economy is fundamentally *different* from our existing economy, in that it is not organised around the prime objective of capitalism: accumulation.

Let me be clear: none of this will be easy. We would be naïve to think otherwise. And there are still difficult questions to which we don't yet have all the answers. No one can give us a simple recipe for a post-capitalist economy; ultimately it has to be a collective project. All I've done here is offer a few possibilities that I hope will nourish the imagination. As for how to make it happen – that will require a movement, as with every struggle for social and ecological justice in history. And to some extent it is already emerging: from the school climate strikes to Extinction Rebellion, from La Via Campesina to Standing Rock; people are not only yearning for a better world, they are mobilising to bring it into being. This sort of thing doesn't just happen on its own. It requires doing the hard work of community organising. The environmentalist movement will need to focus on building alliances with working class and Indigenous formations in order to galvanise a movement capable of capturing political power or forcing incumbents to change course.

I am not a political strategist, but I do want to offer one hopeful observation. Some people worry that there's no way we can possibly accomplish the transition that's required unless we have some kind of totalitarian government impose it from above. But this assumption doesn't hold water. In fact, exactly the opposite is true.

The power of democracy

In 2014, a team of scientists based at Harvard and Yale published a remarkable study on how people make decisions about the natural world. They were interested in whether people will choose to share finite resources with future generations. Future generations pose a problem because they cannot reciprocate with you. If you choose to forgo immediate monetary gain in order to preserve ecology for your grandchildren, they can't offer the favour back – so you gain little from sharing. In light of this, economists expect that people will make a 'rational' choice to exhaust resources in the present and leave future generations with nothing.

But it turns out that people don't actually behave this way. The Harvard-Yale team put people in groups and gave them each a share of common resources to be managed across generations. They found that, on average, a full 68% of individuals chose to use their share sustainably, taking only as much as the pool could regenerate, sacrificing possible profits so that future generations could thrive. In other words, the majority of people behave exactly the opposite to how economic theory predicts.

The problem is that the other 32% chose to liquidate their share of the resources for the sake of quick profits. Over time, this selfish minority ended up depleting the collective pool, leaving each successive generation with a smaller and smaller supply of resources to work with. The losses compounded quickly over time: by the fourth generation the resources were completely exhausted, leaving future generations with nothing – a striking pattern of decline that looks very similar to what's happening to our planet today.

Yet when the groups were asked to make decisions *collectively*, with direct democracy, something remarkable happened. The 68% were able to overrule the selfish minority and keep their destructive impulses in check. In fact, democratic decision-making encouraged the selfish types to vote for more sustainable decisions, because they realised they were all in it together. Over and over again, the scientists found that under democratic conditions, resources were sustained for future generations, at 100% capacity, indefinitely. The scientists ran the experiments for up to twelve generations, and they kept getting the same results: no net depletion. None.[57]

What's so fascinating about this is that it shows widespread and intuitive support for what ecological economists call a 'steady-state' economy. A steady-state economy follows two key principles in order to stay in balance with the living world:

1) Never extract more than ecosystems can regenerate.
2) Never waste or pollute more than ecosystems can safely absorb.

To get to a steady-state economy, we need to have clear caps on resource use and waste. For decades, economists have told us that such caps are impossible, because people will see them as irrational. It turns out they're wrong. If given the chance, this is *exactly* the kind of policy that people want.

*

This helps us see our ecological crisis in a new light. It's not 'human nature' that's the problem here. It's that we have a political system that allows a few people to sabotage our collective future for their own private gain.

How could this be? After all, most of us live in democracies – so why do real-life policy decisions look so different from what the Harvard-Yale experiment predicts? The answer is that our 'democracies' are not actually very democratic at all. As income distribution has grown increasingly unequal, the economic power of the richest has translated directly into increased *political* power. Elites have managed to capture our democratic systems.

We can see this particularly clearly in the United States, where corporations have the right to spend unlimited amounts of money on political advertising, and where there are few restrictions on donations to political parties. These measures – justified according to the principle of 'free speech' – have made it difficult for politicians to win elections without direct support from corporations and billionaires, placing them under pressure to align with the policy preferences of elites. On top of this, large companies and rich individuals spend an extraordinary amount of money lobbying governments. In 2010, $3.55 billion was spent on lobbying, up from $1.45 billion in 1998.[58] And it pays off: one study found that money spent on lobbying the US Congress earned returns of up to 22,000% in the form of tax breaks and profits from preferential treatment.[59]

As a result of political capture, the interests of economic elites in the US almost always prevail in government policy decisions even when the vast majority of citizens disagree with them. In this sense, the US resembles a plutocracy more than a democracy.[60]

Britain exhibits similar tendencies, albeit for different (and older) reasons. Britain's financial hub and economic powerhouse, the City of London, has long been immune from many of the nation's democratic laws and remains free of parliamentary oversight. Voting power in the City of London council is allocated not only to residents, but also to businesses: and the bigger the business, the more votes it gets, with the largest firms getting 79 votes each. In Parliament, the House of Lords is filled not by election but by appointment, with ninety-two seats inherited by aristocratic families, twenty-six set aside for the Church of England, and many others 'sold' to rich individuals in return for large campaign donations.[61]

We can see similar plutocratic tendencies when it comes to finance. A significant chunk of shareholder votes is controlled by massive mutual funds like BlackRock and Vanguard that have no democratic legitimacy. A small number of people decide how to use everyone else's money, and exert extraordinary influence over companies' practices, pushing them to prioritise profits above social and ecological concerns.[62] Then there's the media. In Britain, three companies control over 70% of the newspaper market – and half of that is owned by Rupert Murdoch.[63] In the US, six companies control 90% of all media.[64] It is virtually impossible to have a real, democratic conversation about the economy under these conditions.

The same is true on an international level. Voting power in the World Bank and the IMF – two of the key institutions of global economic governance – is allocated disproportionately to a small number of rich countries. The global South, which has 85% of the world's population, has less than 50% of the vote. Similar problems plague the World Trade Organization, where bargaining power depends on market size. The world's richest economies almost always get their way when it comes to crucial decisions about the rules of the global trade system, while poorer countries – those that have the most to lose from ecological breakdown – are routinely overruled.

One of the reasons we're staring down the barrel of an ecological crisis right now is because our political systems have been completely corrupted. The preferences of the majority who want to sustain our planet's ecology for future generations are trumped by a minority of elites who are quite happy to liquidate everything. If our struggle for a more ecological economy is to succeed, we must seek to expand democracy wherever possible. That means kicking big money out of politics; it means radical media reform; strict campaign finance laws; reversing corporate personhood; dismantling monopolies; shifting to co-operative ownership structures; putting workers on company boards; democratising shareholder votes; democratising institutions of global governance; and managing collective resources as commons wherever possible.[65]

I opened this book by pointing out that large majorities of people across the world are questioning capitalism and yearning for something better. What if we had an open, democratic conversation about what kind of economy we want? What would it look like? How would it distribute resources? Whatever shape it might take, I think it's safe to say it wouldn't look anything like our current system, with its extreme inequality and its tyrannical obsession with endless growth. Nobody actually wants that.

*

We have long been told that capitalism and democracy are part of the same package. But in reality the two may well be incompatible. Capital's obsession with perpetual growth at the expense of the living world runs against the values of sustainability that most of us hold. When people are given a say in the matter, they end up choosing to manage the economy according to steady-state principles that run counter to the growth imperative. In other words, capitalism has a tendency to be anti-democratic, and democracy has a tendency to be anti-capitalist.

This is interesting because both of these traditions emerge, at least in part, from the history of Enlightenment thought. On the one hand the Enlightenment was a quest for the autonomy of reason – the right to question received wisdom handed down by tradition, or by authority figures, or by the gods. This is at the core of how we understand democracy. On the other hand, the dualist philosophy of Enlightenment thinkers like Bacon and Descartes celebrated the conquest of nature as the basic logic of capitalist expansion. Ironically, these two separate projects of the Enlightenment are not allowed to meet. We are not permitted to question capitalism and the conquest of nature. To do so is considered a kind of heresy. In other words, we are encouraged to believe in the values of critical independent thought, but not if it means questioning capitalism.[66]

In an age of ecological breakdown, we must break this barrier down. We must subject capitalism to scrutiny – to reason. The journey to a post-capitalist economy begins with the most basic act of democracy.

Six

Everything is Connected

In the very earliest time
When both people and animals lived on earth
A person could become an animal if they wanted to
And an animal could become a human being.
Sometimes they were people
And sometimes animals
And there was no difference.
All spoke the same language.

Nalungiaq, Inuit elder[1]

We are not the defenders of the river. We *are* the river.

Fisherman, Magdalena River, Colombia

Some images have a way of searing themselves into your mind. I still remember when I first encountered the work of Brazilian photographer Sebastião Salgado. I found myself alone in a dimly lit gallery, face to face with a black-and-white image of a vast desert in Kuwait, a landscape fractured by oil wells, belching thick columns of fire and smoke. And then another: a refugee camp in Tanzania, with makeshift tents sprawling to the horizons, families struggling to survive. And then an open-pit goldmine in the middle of the Amazon rainforest, teeming with men digging shoulder to shoulder, trudging barefoot in the mud, under the watchful eye of armed guards. The images bear witness to the trauma of our civilisation. They haunted me for months.

Salgado spent his career reporting from the front lines of a world in crisis, and eventually it broke him. In the late 1990s, after finishing a project on displacement and migration, he decided to quit photography. 'I was sick. I was not well. I had lost faith in our species,' he told Canada's the *Globe and Mail* newspaper. He and his wife, Lélia, who were living abroad, decided to go back to Brazil. They had inherited his parents' farm, where Salgado spent much of his childhood. He remembered it as a magical forest, a paradise rich with life and flowing with water. But when he returned to the land he found that nothing remained. Intensive livestock farming and deforestation had left it dry, barren and lifeless. The springs had stopped flowing. The hills were eroded. The soil had turned to dust.

As if in a bid to heal a deeper trauma, the Salgados decided to attempt something that everyone told them was impossible – to restore the land to Atlantic rainforest. They began the work in 1999, and the results astonished everyone. Six years later, what had been a 1,730-acre stretch of wasteland was covered over with a layer of hopeful green. And by 2012, the forest had bounced back. The springs were bubbling again, and the animals had returned: birds, mammals, amphibians, even some endangered species. Today that land stands as a beacon of ecosystem restoration, and has inspired many similar projects around the world.

What's powerful about the Salgados' story is that it illustrates how quickly ecosystems can regenerate. The research on this is truly exciting. In 2016, an international team of scientists presented the biggest-ever database on forest regrowth in the New World tropics. They found that across ecosystems, from wet forest to dry forest, it takes an average of only sixty-six years for a forest to recover 90% of its old-growth biomass, completely naturally. All you have to do is leave it alone.[2] Sometimes it happens much faster than this: in Costa Rica, rainforests that had been razed for livestock pasture were found to regrow in as little as twenty-one years, similar to what happened on the Salgados' farm. And while biodiversity generally takes longer to recover, in some cases it can return to old-growth levels in as little as thirty years.[3] As these forests regrow, they pull an extraordinary amount of carbon out of the atmosphere – more than 11 tons of CO_2 per hectare, every year.

These findings offer real hope. It means that if we take the step of scaling down excess industrial activity, the living world can recover with remarkable speed. This is not some distant dream. We would be able to see it happen during our lifetimes, before our very eyes. But we must act quickly, for ecosystems are likely to lose their regenerative capacity as global warming continues.

From this perspective, I cannot help but feel that degrowth is, ultimately, a process of decolonisation. Capitalist growth has always been organised around an expansionary territorial logic. As capital pulls ever-increasing swathes of nature into circuits of accumulation, it colonises lands, forests, seas, even the atmosphere itself. For 500 years, capitalist growth has been a process of enclosure and dispossession. Degrowth represents a reversal of this process. It represents release. It represents an opportunity for healing, recovery and repair.

This is true in a geopolitical sense as well. Remember, excess consumption in high-income nations is sustained by an ongoing process of net appropriation from the lands and peoples of the global South, on unequal terms. Colonialism as such may have ended half a century ago, at least in most parts of the world, but – as we've seen – those old patterns of plunder continue to this day, with ruinous consequences. To the extent that degrowth in high-income nations releases global South communities from the grip of extractivism, it represents decolonisation in the truest sense of the term. Indeed, degrowth demands have been articulated by social movements in the global South for some time, *avant la lettre*. We can see this in the People's Agreement of Cochabamba, drafted in 2010 by grassroots organisations from more than a hundred and thirty countries, which recognises the colonial dimensions of growthism in the North and calls for rich countries to reduce their use of the planet's resources and end their exploitation of poorer countries.

*

My years of researching degrowth have given me something I didn't really expect – hope. And yet I have nonetheless found myself worrying, from time to time, that something is still missing. By focusing all our attention on how to fix the economy, we risk ignoring the bigger picture. Yes, we must take steps to evolve beyond capitalism. But capitalism is only the proximate driver of the crisis we face; it's not really the underlying condition. That's something that lies much deeper.

Remember, the rise of capitalism in the sixteenth and seventeenth centuries didn't come out of nowhere. As we saw in Chapter 1, it required violence and dispossession and enslavement; but even more than that it required crafting a new story about nature. It required getting people to see nature, for the first time, as something fundamentally distinct from humans; something not only inferior and subordinate, but devoid of the animating spirit we ascribe to people. It required splitting the world in two. It required, in a word, separation. For the past 500 years, the dominant culture on our planet – the culture of capitalism – has been rooted in that rift.

Once we grasp this, then it becomes clear that the struggle before us is more than just a struggle over economics. It is a struggle over our very theory of being. It requires decolonising not only lands and forests and peoples, but decolonising our minds. To begin this journey, we need new sources of hope, new well-springs of possibility – new visions for how things could be otherwise. What we will learn along the way is that the secret to building an ecological civilisation isn't at all about limits and meagreness. It is about something radically bigger. Bigger than we can imagine.

Lessons from the ancestors

One of the real pleasures I've discovered in my career as an anthropologist has been the process of piecing together a much deeper sense of the human story than I used to have. I remember, as a postgraduate student, walking out of classes sometimes feeling almost overwhelmed with a sense of new perspective, as if I had just stepped out of a prosaic little cottage only to find myself on the lip of a vast escarpment, with landscapes of time rolling out before me. The story of humanity plays out like a journey, with our ancestors venturing out of Africa and migrating across the planet, over tens of thousands of years. Along the way they encountered a vast array of different ecosystems – from savannah to deserts, jungles to steppes, wetlands to tundra. With each new zone they entered, they had to learn how these ecosystems worked so that they could live within them sustainably, in reciprocity with the other species they depended on for nourishment and sustenance. Sometimes they succeeded. Sometimes they failed.

Nowhere was this mixed record more pronounced than during the Austronesian expansion, when humans left mainland Asia about two or three thousand years ago and settled throughout the vast network of islands that stretches south and east into the Pacific Ocean. The people who set out on these expeditions came from a culture that was established in the crucible of an enormous continent, governed by stable monsoon weather conditions, where they regularly terraformed whole river basins for agriculture. Living on such a vast expanse of territory, they were accustomed to feeling like they had seemingly endless resources at their disposal – like they could do whatever they wanted to the land.

The migrants took this culture with them when they landed on the islands of Austronesia. But the expansive logic of the mainland civilisations didn't work out quite so well on the islands. In fact, the consequences were devastating. Settlers tore through island megafauna in a bonanza of easy protein – giant turtles, birds, fish and other easy prey that were unaccustomed to human predators. They chopped down trees to clear the land for crops. All of this might have had little consequence on the mainland, but on the islands it proved to be disastrous. Keystone species died off. Ecosystems fell out of balance. Life began to unravel. A number of societies completely collapsed. Some of the islands were abandoned altogether.

But as the Austronesian expansion wore on, settlers learned from their mistakes. They learned that to build a thriving society within a bounded island ecosystem requires a completely different approach to ecology. They had to swap the ideology of expansion for an ideology of integration. They had to learn to pay attention to other species – learn their habits, their languages, and their relationships with others. They had to learn how much they could safely take from any given community, and how to give back in order to ensure its continuation. They had to learn not only to protect but to *enrich* the island ecosystems on which they depended. They had to develop new, more ecological ways of thinking about their relationship with animals and forests and rivers, and they had to build these into their myths and rituals so they would never be forgotten. Societies that took these steps ended up thriving in the Pacific islands.

Today we stand at a similar juncture, and the future could go either way. We are a civilisation obsessed with expansion that has suddenly discovered, as it were, that it inhabits an island. Will we cling to our reckless old ideologies, or will we seek to learn a new, more intelligent way of being? Fortunately, if we choose the latter course, we do not have to start from scratch. Humans have developed ways of being ecological in an astounding variety of places. If we look to communities that live close to the land today, we can find a wealth of clues about what real ecological intelligence looks like.

On being ecological

If you've ever seen photographs from inside the Amazon rain-forest, you'll have had a glimpse of what it's like there. Dense, steamy, tangled and teeming with life. It's also home to hundreds of Indigenous communities who have inhabited the region for many generations, including – along the invisible border between Ecuador and Peru – one group known as the Achuar.

Over the past decade or two, the Achuar have attracted attention because there's something rather unexpected about their world view that has riveted anthropologists and philosophers, and it's now completely upending the way that they think about nature. For the Achuar, you see, 'nature' does not exist. This might seem absurd to Western observers, who tend to see the category of nature as self-evident. It certainly seemed absurd to me when I first encountered it. But linger with this idea for long enough and it becomes clear there's something profound going on. And it may hold powerful secrets within.

If you visit the Achuar, you'll find them living in small circular clearings in the middle of jungle, with dense walls of trees rising up all around them like giant waves of green – dark, brooding, pulsing with the noises of frogs, toucans, snakes, monkeys, jag-uars, millions upon millions of insects, plus a universe of mosses and mushrooms and curling, roping vines. For many people, to live in such a way, cut off from other human communities, would feel tremendously lonely and isolated. But the Achuar see the jungle quite differently. They see people all around.

As far as the Achuar are concerned, most of the plants and ani-mals that populate the jungle have souls (*wakan*) similar to the souls of humans, and are therefore classified, literally, as 'per-sons' (*aents*). Just like humans, plants and animals have agency, intentionality and even self-consciousness. They experience emotions and exchange messages, not only among themselves but also with other species, and even – through dreams – with humans. There is nothing that fundamentally distinguishes them, in essence, from people. In fact, the Achuar go so far as to regard plants and animals as their *relatives*. The monkeys and other animals they hunt for food are regarded as brothers-in-law, and the relationship between them is governed by similar rules of circumspection and mutual respect. As for the plants they rely on for food, they are regarded as children to be nourished and cared for. For the Achuar, the jungle is not just a source of sustenance. It is a terrain full of intimate connections and kinship.

It might be tempting to dismiss all of this as nothing but quaint metaphor. But it's not. Just as we know that maintaining good relations with our partners and children and in-laws and neighbours is essential to maintaining a secure, happy life, so the Achuar know that their existence depends on maintaining good relationships with the teeming community of non-human (or more-than-human) persons with whom they share the forest. They know that they are fundamentally interdependent; that without them they would be nothing – non-existent. Their fates are bound together.

These same principles are held by most of the peoples that inhabit the Amazon rainforest. It is a widespread and completely normal way of interacting with the world. But it is not just Indigenous Amazonians who hold these views. This ethic is widely shared – albeit with important variations – among countless Indigenous communities on every continent.[4] It is remarkable in its consistency. And in many cases it's not only plants and animals that are regarded as persons, but also inanimate beings like rivers and mountains.

Take the Chewong, for example – the Indigenous community that inhabits the tropical forests of the Malay peninsula, on the other side of the planet from the Amazon. While their population barely reaches 300, they say that their community extends far beyond humans to encompass the plants, animals and rivers of the forest. In fact, they go so far as to refer to them collectively as 'our people' (*bi he*). Once again, this is not simply a romantic metaphor. The Chewong regard all beings as underpinned by the same moral consciousness (*ruwai*). Squirrels and vines and humans may appear to be radically different on the face of it, but beneath this veneer all ultimately participate in the same moral being. As such, all have an ethical responsibility to ensure that the broader, collective ecological system runs smoothly, maintaining the intimate interdependencies that constitute the web of life. Bees are morally responsible for the welfare of humans just as humans are responsible for the welfare of bees.

Four thousand kilometres away, on the island of New Guinea, the Bedamuni people have a saying: 'When we see animals, we might think that they are just animals, but we know that they are really like human beings.' The Kanaks on nearby New Caledonia island have a similar ethic, which they extend not only to animals but to plants as well. They insist that there is a material continuity between humans and plants: humans and plants have the same kinds of bodies, they say – to the point where ancestors will return to inhabit certain trees after passing away. The Bedamuni and the Kanaks reject the formal distinctions between humans, plants and animals that Westerners tend to take for granted, and they refuse to accept any hierarchies among them. There's nothing at all like the Great Chain of Being that has sat at the heart of Western philosophy for so long, with humans at the top and everything else staggered out below.

For these communities, it is impossible to draw distinctions between humans and 'nature', as those of us who live in capitalist societies so routinely do – a legacy handed down to us by early Mesopotamian civilisations, transcendental religions and Enlightenment philosophers like Bacon and Descartes. Such a distinction would make no sense. Indeed, it would be morally reprehensible, almost even violent. It would be like one group of people denying the humanity of another group, seeking to exclude them from rights on racist grounds – just as Europeans once did in order to justify colonisation and slavery. It would seem like an affront to the right way of living, which requires an understanding of interdependence.

*

This way of seeing the world has powerful implications for how people interact with their ecology. What do you do with a natural world that is infused with the very same kind of personhood that humans have? With beings that are regarded as living in social community alongside humans, even in the role of relatives? It is unthinkable to regard such beings as 'natural resources', or as 'raw materials', or even as 'the environment'. From the perspective of the Achuar, the Chewong and other Indigenous groups, to see nature as a resource and to exploit it is ethically unfathomable. After all, to exploit something you must first regard it as less than human – as an object. This is impossible in a world where nothing is less than human, and where all beings are subjects in their own right.

Don't get me wrong. Obviously these communities take from their surrounding ecology. They fish, they hunt, they grow orchards that provide them with fruits and nuts and tubers to eat. And indeed this presents a question. For if animals are persons, then eating them would seem to be a form of cannibalism. As one Arctic shaman put it to the anthropologist Knud Rasmussen, 'The greatest peril of life lies in the fact that human food consists entirely of souls.'

This seems like an impossible conundrum; but it is impossible only to those who insist on the separation between humans and non-humans in the first place. If you start from the premise that both parties are elements of the same whole, the conundrum melts away. What matters is not one or the other, but the relationship. Suddenly it becomes a question of equilibrium and balance. Yes, humans hunt toucans and dig up tubers, but when they engage in these activities they do so in the spirit not of extraction but of *exchange*. It is a matter of mutual reciprocity. The moral code at play here is not that you should never take (that would lead to a quick demise), but that you should never take more than the other is willing or able to give – in other words, never more than an ecosystem can regenerate. And you have to make sure to give back in return, by doing what you can to enrich, rather than degrade, the ecosystems on which you depend.

This takes a lot of work. It requires listening, empathy, dialogue. For many Indigenous communities, the skills of managing relations between human and non-human beings are honed in particular by shamans. For much of the twentieth century, anthropologists believed that the shaman's role was limited to serving as a medium between humans and their ancestors. Now it's increasingly clear that in many cases shamans also mediate between the human community and the broader community of beings on which humans depend.

Shamans grow to know these other beings intimately. In the Amazon, they communicate with them in trances and dreams, transmitting messages and intentions back and forth. Because shamans spend so much time interacting with their non-human neighbours, they have an expert's grasp on how ecological systems work. They know exactly how many fish – and of what species – can be taken from a river in any given season while ensuring that plenty are able to spawn for the next year. They know how many monkeys can be safely hunted without harming a troupe. They know when a grove of fruit trees is healthy, and when it's in trouble. They use this knowledge to make sure that humans never take more from their plant and animal relatives than the forest can safely provide.

In this sense, the shaman operates as a kind of ecologist; an expert who understands and maintains the fragile interdependencies that constitute the jungle ecosystem, with knowledge of botany and biology that may far outstrip that which even the most prestigious university professors would dare to claim.

*

What a thrilling way to experience the world! For those of us raised in capitalist culture, trained in the conceits of dominion and dualism, it is almost impossible to comprehend. How much richer would our experience of the living world be if we regarded it as pulsing with intention and sociality? Who lives there? What are they like? What is their experience? What will we say to one another? Even just to imagine living this way seems like a portal to an enchanted world – one that's hidden somehow right in plain view.

Anthropologists refer to this way of being as animism. The religious studies scholar Graham Harvey defines animism quite simply as the claim 'that the world is full of persons, only some of whom are human, and that life is always lived in relationship with others'.[5] Animists approach animals and plants and even rivers and mountains as subjects in their own right, rather than as objects. There is no 'it' in such a world view. Everything is 'thou'.[6]

This is the key bit to understand. Some people make the mistake of thinking that when animists talk about non-human beings as 'persons', they are merely projecting human qualities onto them, seeing them (mistakenly) as humans in disguise. But that's not what's going on here. Rather, animists recognise other species as *subjects* – subjects who have their own subjective, sensory experience of the world, just as we humans do. And it is precisely because they are subjects that they are regarded as persons. Because to be a subject *is to be a person*.

It's not difficult to imagine how people might arrive at this conclusion. Indigenous communities that depend on foraging and hunting in the forests have to get to know their local plants and animals intimately. They spend tens of thousands of hours learning and imitating the calls of monkeys and birds and jaguars, to the point of mastering subtle differences in meaning and mood – skills that are essential for successful hunting. They will get to know the preferences of various plants for different kinds of soils, how they move in response to changes in temperature and light, and how they interact with beetles and ants and birds. Their lives depend on mastering this kind of knowledge. And in the process, they come to realise – how could they not? – that all these beings are experiencing the world in their own ways, with their own unique set of senses, and interacting and responding to it with their own type of intelligence. It is a process of radical empathy with non-human persons.[7]

It seems obvious, in some ways. And yet it's all too easy for us to forget – particularly if we live in cities, where people rarely if ever encounter other species as anything but decoration. Even in rural areas, on farms, wild species are quite often treated as mere pests, to be exterminated if at all possible. In these contexts, we easily slip into thinking of other beings not as subjects but as objects – when we think of them at all. Or maybe it's not that we forget, or that we slip ... maybe it's that we subconsciously suppress what we know on some deep level to be true, because to let ourselves think about the fact that our economic system depends on the systematic exploitation of other living beings is just too much to bear.

Whatever one might think of animism, one thing is certain: it is deeply ecological. In fact, it anticipates the core principles of ecological science that lie at the heart of the discipline today, which can be boiled down into a single phrase: everything is intimately interconnected; behave accordingly. And this is not just nice rhetoric. It works. Living this way has real, material effects on the world. Scientists estimate that 80% of the planet's biodiversity is to be found on territories stewarded by Indigenous peoples.[8] Clearly they are doing something right. They've protected life. They've nourished it. Not out of charity, or because it's beautiful, but because they recognise the fundamental interdependence of all beings.

As growthism accelerates the sixth mass extinction event in our planet's history, the contrast between animist values and capitalist values could hardly be more pronounced.

Minority reports

To people unfamiliar with these ideas, animism may seem a bit strange at first, possibly even bizarre. This is not surprising. After all, we are heirs of René Descartes and the dualist philosophy that came to define the Enlightenment, which proceeds from exactly the opposite premise.

Remember, Descartes started with the old monotheistic idea of a fundamental distinction between God and creation, and then took it one step further. Creation itself is divided into two substances, Descartes said: with mind (or soul), on the one hand, and mere matter on the other. Mind is special; it is part of God. It cannot be described with the normal laws of physics or maths. It is an ethereal, divine substance. Humans are unique among all creatures in having minds and souls, which is the mark of their special connection to God. As for the rest of creation – including the human body itself – it is nothing but inert, unthinking matter. It is but 'nature'.

Descartes' ideas had no grounding in empirical evidence, but they became popular among European elites in the 1600s because they bolstered the power of the Church, justified the capitalist exploitation of labour and nature, and gave moral licence to colonisation. Even the very idea of 'reason' itself came to rely on these assumptions. Humans alone have reason, Descartes argued, because we alone have minds. And the first step of reason is to realise that we – our minds – are *separate from* our bodies, and separate from the rest of the world.

From this perspective, the animist insistence on seeing the world as intimately interconnected was long regarded as irrational and unenlightened. In the nineteenth century, prominent anthropologists described it as 'childish': only children see the world as enchanted, but this is a cognitive error that we must correct. Indeed, not only reason but modernity itself – and modern science – came to be defined with respect to categorical distinctions between humans and nature, subject and object. Animism provided the perfect foil for the emerging concept of the 'modern'.

But Descartes didn't have the last word. Even as the ink was drying on his manuscripts, he came under attack from his own contemporaries who pointed out fundamental errors in his work. And over the 400 years since then, advances in scientific research have proved not only that Descartes was wrong, but that animist thought is in key respects more resonant with how life and matter actually work.

*

The backlash against Descartes started with a brave Dutch philosopher by the name of Baruch Spinoza. Spinoza grew up in a Sephardic Jewish family in Amsterdam in the 1600s, just as Descartes was becoming a celebrity. But while the elites of the day fawned over Descartes' dualism, Spinoza wasn't convinced.

In fact, he took exactly the opposite view. Spinoza pointed out that the universe must emerge from one ultimate cause – what today we might recognise as the Big Bang. Once we accept this fact, Spinoza argued, then we have to accept that while God and souls and humans and nature might seem to be fundamentally different kinds of entities, they are really just different aspects of a single, grand Reality – a single substance – and governed by the same forces. This has radical implications for the way we think about the world. It means that God ultimately participates in the same substance as 'creation'. It means that humans participate in the same substance as nature. It means that mind and soul are the same substance as matter. In fact, it means that everything is matter, everything is mind, and everything is God.

These ideas were heretical at the time. No souls? No transcendental God? Spinoza's teachings upended the core tenets of religious doctrine, and threatened to pry open difficult moral questions about the exploitation of nature and labour. After all, if nature is ultimately the same substance as God, then humans can hardly claim dominion over it.[9]

The backlash was swift and severe. Spinoza ran so against the grain of establishment thinking at the time that he found himself at the sharp end of brutal persecution. The Jewish authorities in Amsterdam issued a *herem* against him, expelling him from the community. The Christian establishment threw him out too; and the Catholic Church went so far as to list his works in the *Index of Forbidden Books*. His own family shunned him, and he suffered physical assault on the streets. At one point he was stabbed on the steps of a synagogue by an assailant shouting 'Heretic!'. But none of this deterred him. Spinoza kept the torn cloak he was wearing when stabbed, and wore it as a symbol of defiance.

*

Europe faced a fork in the road. They had two options: the path of Descartes or the path of Spinoza. With the full backing of the Church and capital, Descartes' vision won out. It gave legitimacy to the dominant class forces, and justified what they were doing to the world. As a result, today we live in a culture shaped by dualist assumptions. But it could have been otherwise. I often find myself wondering how things might have turned out differently if Spinoza's perspective had prevailed instead. How would this have shaped our ethics? Our economics? Perhaps we wouldn't now be facing the nightmare of ecological collapse.

What's so striking about this story is that, over the centuries that followed, scientists affirmed a number of Spinoza's claims. They affirmed that there is in fact no fundamental distinction between mind and matter; that mind is an assemblage of matter, just like everything else. They affirmed that there is no fundamental distinction between humans and non-human beings; that humans and non-humans evolved from the same predecessor organisms. And they affirmed that everything in the universe is ultimately governed by the same physics – even if that physics has yet to be fully described. Ironically for a school of thought that was once considered the height of Enlightenment science, dualism ended up suffering a tremendous defeat at the hands of science itself. Indeed, today the tables have turned: Spinoza is now routinely recognised as one of the best thinkers in modern European philosophy, and celebrated as a key figure in the history of science.

And yet even as science broke from dualism, some of Descartes' assumptions about the world lingered on. To this day, most people in Western societies still believe that humans are fundamentally set apart from the rest of nature. To justify this belief, religious people might fall back on some notion of the soul. Atheists, for their part, will insist that it has something to do with intelligence, or consciousness. Only humans, they'll say, have an inner self, and the capacity to reflect on the world – and this is what makes us superior to other beings. Only humans are real *subjects*, while other beings are 'objects' in our field, mechanically playing out their lives according to genetic codes. Four hundred years later and we're still retweeting Descartes.

Beginning in the middle of the twentieth century, philosophers like Edmund Husserl and Maurice Merleau-Ponty began to question these everyday assumptions using a new framework called phenomenology. They pointed out that human consciousness, and therefore the self, cannot exist in some abstract, transcendental mind. All consciousness is derived from the experience of phenomena, and experience fundamentally depends on the body. Everything we know, everything we think – indeed, our very sense of self – derives from our embodied experience in the world. The philosopher David Abram puts it in these poetic words:

> Without this body, without this tongue or these ears, you could neither speak nor hear another's voice. Nor could you have anything to speak about, or even to reflect on, or to think, since without any contact, any encounter, without any glimmer of sensory experience, there could be nothing to question or to know. The living body is thus the very possibility of contact, not just with others but with oneself – the very possibility of reflection, of thought, of knowledge.[10]

Of course, none of this was particularly surprising to people who were already all too aware of their bodies – people, and particularly women, who depend on sometimes painful manual labour for their livelihood, be it in the fields or the factory or the home. But the rise of phenomenology marked the moment that Europe's elite men discovered that they had bodies; that they were not just reason in a vat. It collapsed the mind-body distinction once and for all.

Once you accept this, it's a short step to recognising that those other 'phenomena' that populate our field of experience, the other beings with whom we engage – not just other humans but plants and animals as well – they are beings with subjective experience too. After all, they are bodies, like us, sensing the world, engaging with it, responding to it, shaping it. In fact, the world that presents itself to us is co-created by other subjects, just as we co-create their world. We are all engaged with each other in a sensual dance of perception, an ongoing dialogue through which we come to know the world.

When we think of it this way, suddenly the subject-object distinction collapses. Husserl argued that the universe of experience isn't defined by subject-object relations; rather, it is an inter-subjective field which is collectively produced. Everything we know, everything we think, everything we *are*, is shaped by mutual interaction with other subjects.

These insights from phenomenology bring us remarkably close to what animists have so long insisted upon. After all, if we start from the belief that what makes humans special is the fact that they are *subjects*, then once we realise that non-human beings are subjects too we're in completely new terrain. Suddenly the boundaries of personhood stretch out well beyond the human community to encompass non-human others.

*

I've mentioned Western thinkers here simply to show that there have always been minority reports even within Western philosophy itself. But these ideas have been developed, practised and kept alive most fully by Indigenous thinkers who have not been encumbered by Cartesian assumptions in the first place, such as the Honduran activist Berta Cáceres, who was assassinated in 2016 for defending the Rio Gualcarque; the Inuit leader Sheila Watt-Cloutier, who was nominated for the Nobel Peace Prize in 2007 before it went to Al Gore; the Brazilian Indigenous activist and leader Ailton Krenak; and, to mention two people who have been particularly influential to me, the Algonquian scholar-activist Jack D. Forbes and the Potawatomi scientist and philosopher Robin Wall Kimmerer.

Reading these remarkable people always reminds me of Aimé Césaire's words, which I mentioned towards the beginning of this book. Remember, Césaire described colonisation as a process of 'thingification'. Living beings, nature and humans alike, had to be rendered as objects so they could be legitimately exploited. This paved the way for cheap nature and capitalist growth. Given this history, it becomes clear that any process of decolonisation must therefore begin with a process of *dethingification*. This is what Indigenous scholars teach us: that we must learn to see ourselves once again as part of a broader community of living beings. If our approach to degrowth does not have this ethic at its heart, then we have missed the point.

A second Scientific Revolution

In the late twentieth century, phenomenology managed to re-implant animist principles right into the heart of European philosophy. And science quickly followed. Over the past two decades, a cascade of scientific discoveries has started to radically change how we perceive ourselves in relation to the rest of the living world.

Take bacteria, for instance. For generations we were told that bacteria were bad – full stop. We armed ourselves with anti-bacterial soaps and chemical disinfectants and set out to purify our bodies and our homes and even our food of the invisible little enemies we call germs. But in recent years scientists have overthrown many of those early misconceptions.

Our gut, skin and other organs are populated by trillions of microbial beings – and it turns out that we depend on these little creatures for our very existence. Gut bacteria are vital for digestion, as they break down food and turn it into nutrients we can use. They help regulate our immune responses. They are even essential to healthy brain function, as they activate neural pathways and nervous-system signalling mechanisms that help us deal with stress, prevent anxiety and depression, and promote mental health. They may even play a role in our social lives: scientists have recently discovered that wiping out the microbiota in mice makes them behave in antisocial ways, and they anticipate that the same is likely to be true of humans.[11] These facts utterly confound any clear distinctions between mind and body, human and 'nature'. The assumptions that underpin dualist thought are disintegrating in the face of science.[12]

And it's not just bacteria. Even some viruses appear to be beneficial to us, like the phages that regulate bacterial populations.[13]

Without them, bacterial processes in our bodies could tip out of balance.

If you were to count up all the cells that constitute your body, you'd find that more of them belong to other lifeforms than belong to 'you' as such.[14] Let this fact sink in, and it upends the way we think about ourselves. What is a self, anyway, if it cannot readily be distinguished from the trillions of other beings with whom we live, with whom we co-manage our physical and mental states, and without whom we cannot survive? As the British philosopher of science John Dupré has put it, 'These findings make it hard to claim that a creature is self-sufficient, or even that you can mark out where *it* ends and another one begins.'[15]

Things get even stranger when we zoom out over evolutionary time. Humans have two sets of DNA – one contained in the nucleus of each of our cells, and the other in the mitochondria, an 'organelle' that lives within the cell itself. Biologists believe that this second set, the mitochondrial DNA, is derived from bacteria that were engulfed by our cells at some point in the evolutionary past. Today these little organelles play an absolutely essential role in human life: they convert food into energy that our bodies can use. This is mind-bending: that our most basic metabolic functions, and even the genetic codes that constitute the very core of who we are, depend on other beings.

The implications of this are profound. A team of scientists associated with the Interdisciplinary Microbiome Project at Oxford University have suggested that discoveries related to bacteria may revolutionise not only our science but our ontology too: 'Our ability to map previously invisible forms of microbial life in and around us is forcing us to rethink the biological constitution of the world, and the position of humans vis-a-vis other forms of life.'

*

Just as bacteria are revolutionising how we think about our relationship with the world, biologists are also discovering some remarkable things about trees and forests that are upending how we think about plants.

When we see a tree, we tend to think of it as a singular unit – just as we think of ourselves as individuals. But biologists have discovered that it's not quite so simple. They have come to understand that trees depend on certain kinds of fungi in the soil: hair-thin structures called hyphae that interlace with cells in the roots of trees to form mycorrhiza. The fungi benefit by receiving some of the sugar that plants produce through photosynthesis (which it cannot otherwise make), while the trees benefit in turn by receiving elements like phosphorous and nitrogen that they cannot produce for themselves, and without which they cannot survive.

But this reciprocity is not confined to just two parties in this ancient relationship. Invisible fungal networks also connect the roots of different trees to one another, sometimes over great distances, forming an underground internet that allows them to communicate, and even to share energy, nutrients and medicine. The ecologist Robert Macfarlane explains how this works:

> A dying tree might divest itself of its resources to the benefit of the community, for example, or a young seedling in a heavily shaded understory might be supported with extra resources by its stronger neighbours. Even more remarkably, the network also allows plants to send one another

warnings. A plant under attack from aphids can indicate to a nearby plant that it should raise its defensive response before the aphids reach it. It has been known for some time that plants communicate above ground in comparable ways, by means of airborne hormones. But such warnings are more precise in terms of source and recipient when sent by means of the myco-net.[16]

Trees co-operate. They communicate. They share. Not only among members of the same species, but across species barriers: Douglas firs and birches feed each other. And it's not just trees; we now know that all plants – except for a handful of species – have this same relationship with mycorrhiza. Just as with our gut bacteria, these findings challenge how we think about the boundaries between species. Is a tree really an individual? Can it really be conceived as a separate unit? Or is it an aspect of a broader, multi-species organism?

There's also something else going on here – something perhaps even more revolutionary. Dr Suzanne Simard, a professor in the department of forest & conservation at the University of British Columbia, has argued that mycorrhizal networks among plants operate like neural networks in humans and other animals; they function in remarkably similar ways, passing information between nodes. And just as the structure of neural networks enables cognition and intelligence in animals, mycorrhizal networks provide similar capacities to plants. Recent research shows that the network not only facilitates transmission, communication and co-operation – just like our neurons do – it also facilitates problem-solving, learning, memory and decision-making.[17]

These words are not just metaphorical. The ecologist Monica Gagliano has published groundbreaking research on plant intelligence, showing that plants remember things that happen to them, and change their behaviour accordingly. In other words, they *learn*. In a recent interview with *Forbes*, she insisted: 'My work is not about metaphors at all; when I talk about learning, I *mean* learning. When I talk about memory, I *mean* memory.'[18]

Indeed, plants actively change their behaviour as they encounter new challenges and receive messages about the changing world around them. Plants sense: they see, hear, feel and smell, and they respond accordingly.[19] If you've ever seen time-lapse footage of a vine growing up a tree, you'll have an idea of what this looks like in action: that vine is no automaton – it's sensing, moving, balancing, solving problems, trying to figure out how to navigate new terrain.

The more we learn, the stranger (or perhaps more familiar?) it all becomes. Simard's work shows that trees can recognise their own relatives through mycorrhizal networks. Older 'mother' trees can identify nearby saplings that came from their own seeds, and they use this information to decide how to allocate resources in times of stress. Simard also describes how trees seem to have 'emotional' responses to trauma in a way that's not dissimilar to animals. After a machete whack or during an aphid attack, their serotonin levels change (yes, they have serotonin, along with a number of neurochemicals that are common in animal nervous systems), and they start pumping out emergency messages to their neighbours.

Of course, none of this is to say that plant intelligence is exactly like that of animals. In fact, scientists warn that our urge to constantly compare the intelligence of some species with that of others is exactly the problem: it ends up blinding us to how other kinds of intelligence might work. Set out in search of a brain and you'll never even notice the mycorrhiza that have been pulsing through the earth, evolving right under our feet, for 450 million years.

This research is just taking off, and we have no idea where it might lead. But Simard is careful to point out that it's not exactly new:

> If you listen to some of the early teachings of the Coast Salish and the Indigenous people along the western coast of North America, they knew [about these insights] already. It's in the writings and in the oral history. The idea of the mother tree has long been there. The fungal networks, the below-ground networks that keep the whole forest healthy and alive, that's also there. That these plants interact and communicate with each other, that's all there. They used to call the trees the tree people . . . Western science shut that down for a while and now we're getting back to it.[20]

*

Trees aren't only connected with each other. They are also connected with *us*. Over the past few years, research into human–tree relationships has yielded some truly striking findings.

A team of scientists in Japan conducted an experiment with hundreds of people around the country. They asked half of the participants to walk for fifteen minutes through a forest, and the other half to walk through an urban setting, and then they tested their emotional states. In every case, the forest walkers experienced significant mood improvements when compared to the urban walkers, plus a decline in tension, anxiety, anger, hostility, depression and fatigue.[21] The benefits were immediate and effective.

Trees also have an impact on our behaviour. Researchers have found that spending time around trees makes people more co-operative, kinder and more generous. It increases our sense of awe and wonder at the world, which in turn changes how we interact with others. It reduces aggression and incivility. Studies in Chicago, Baltimore and Vancouver have all discovered that neighbourhoods with higher tree cover have significantly fewer crimes, including assault, robbery and drug use – even when controlling for socio-economic status and other confounding factors.[22] It's almost as though being with trees makes us more *human*.

We don't know quite why this happens. Is it just that green environments are somehow more pleasant and calming? A study in Poland suggests that doesn't explain it. They had people spend fifteen minutes standing in a wintertime urban forest: no leaves, no green, no shrubbery; just straight, bare trees. One might think such an environment would have minimal if any positive impact on people's mood, but not so: participants standing in the bare forest reported significant improvements in their psychological and emotional states when compared to a control group that spent those fifteen minutes hanging out in an urban landscape.[23]

And it's not just mood and behaviour. It turns out that trees have an impact on our physical health too – in concrete, material terms. Living near trees has been found to reduce cardiovascular risk.[24] Walking in forests has been found to lower blood pressure, cortisol levels, pulse rates and other indicators of stress and anxiety.[25] Even more intriguingly, a team of scientists in China found that elderly patients with chronic health conditions demonstrated significant improvements in immune function after spending time in forests.[26] We don't know for sure, but this may have something to do with the chemical compounds that trees exhale into the air. The aromatic vapours released by cypress, for example, have been found to enhance the activity of a number of human immune cells, while reducing stress hormone levels.[27]

In an attempt to quantify the overall benefit of trees, scientists in Canada found that trees have a more powerful impact on our health and well-being than even large sums of money. Having just ten more trees on a city block decreases cardio-metabolic conditions in ways comparable to earning an extra $20,000. And it improves one's sense of well-being as much as earning an extra $10,000, moving to a neighbourhood with $10,000 higher median income, or *being seven years younger.*[28]

These results are astonishing. There's a real mystery here, which scientists still do not yet understand. But perhaps we shouldn't be so surprised. After all, we have co-evolved with trees for millions of years. We even share DNA with trees. After countless generations, we've come to depend on them for our health and happiness just as we depend on other humans. We are, in a very real sense, *relatives*.

*

These remarkable interdependencies – trees, fungi, humans and bacteria – are only the very tip of the iceberg. Ecologists are finding them literally everywhere. There is not a single ecosystem on the planet where species don't interact in mutually enriching ways. We are even starting to rethink the relationship between predators and their prey. In the past we saw this as a matter of domination and plunder – 'dog eat dog', 'the law of the jungle', 'kill or be killed'. And certainly if you zoom in on discrete moments of predation they can be quite gruesome, as you'll know if you've ever seen footage of a lion on the hunt. But zoom out and it becomes clear that there's something else going on. Predation turns out to be more about balance and equilibrium than anything else.

In Alaska, for example, wolves keep caribou populations in check. This prevents the caribou from overgrazing saplings, which in turn allows forests to grow and flourish. Forests prevent erosion, which keeps soils healthy and enables rivers to run clear. Good soils give rise to berries and grubs, while clear rivers provide habitats for fish and other freshwater creatures. Fish and berries and grubs in turn feed bears and eagles. These interdependencies build strength and resilience into ecosystems, literally fleshing out the network. But the cascades of generosity also work in reverse. In areas where wolves have been exterminated, whole ecosystems fall apart: forests collapse, soils erode, rivers fill with silt, and eagles and bears disappear.

Similar ecosystem dynamics have been described in every region of every continent, including at the poles. Nothing exists alone. Individuality is an illusion. Life on this planet is an interwoven mesh of relational becoming.

There is even evidence that these principles operate at a planetary level, between entire Earth-systems processes. Scientists have been learning how plant, animal and bacterial biomes interact with the land, the atmosphere and the oceans in ways that regulate everything from the temperature of the planet's surface to the salinity of the seas to the composition of the air. Our planet is one, giant system of interlocking reciprocities. The British scientist James Lovelock and his American collaborator Lynn Margulis have described the Earth as a superorganism, which automatically self-regulates in a manner that maintains the conditions for life, just as the human body self-regulates to keep internal systems in functional balance. This is the Gaia hypothesis, so named after the goddess of the Earth in Greek mythology. And indeed these findings from Earth-systems science and biogeochemistry would not be surprising to peoples who have long regarded the Earth as a living being, or even as a mother.

Post-capitalist ethics

What does all of this mean for us? How should we live in the light of this science?

Let's go back to those findings about plants, just for the sake of argument. When research about plant intelligence first began circulating on social media, not everyone reacted well to it. If plants are intelligent, perhaps even conscious in some distributed sense, then how are we supposed to deal with the fact that harvesting crops must therefore be a kind of murder? How are we supposed to cut trees for furniture if it means splitting up a family? Thinking this way would make life so ethically fraught as to be practically impossible. For many people, this conundrum poses such a problem that they feel the only reasonable response is to reject the science itself.

Interestingly, these are the very dilemmas that the Achuar, Chewong and other animist communities face. And perhaps we can take lessons from the answers they've arrived at after generations of thinking about it. There's nothing *necessarily* unethical about harvesting crops or cutting down trees, they say – or even hunting and eating animals, for that matter. What's unethical is to do so without gratitude, and without reciprocity. What's unethical is to take more than you need, and more than you give back. What's unethical is exploitation, extraction and, perhaps worse still, waste.

Remember, for the Achuar and Chewong, the key principle is reciprocity. You have to start by recognising that you are in a relationship of interdependence. Robin Wall Kimmerer argues that the ethics of this exchange must begin from the awareness that we are engaging with sovereign beings. It is a relationship with persons who are deserving of our respect. Kimmerer points out that we should receive food and materials from the living world with the same care and decorum and gratitude that we might receive a healthy, home-cooked meal from our grandmother. We should treat what we receive not as a right, but as a gift.[29]

This is not just about uttering a 'Thank you' beneath our breath and moving on with our lives (although practising even this simple act can completely change how we interact with the living world). It is much more than that. What's powerful about gifts is that they place us in a position of self-restraint, where we are careful to take no more than we need, and no more than the other is able to share. This has intrinsic conservational value, and it's a radical act in the context of a culture that's hell-bent on consumption far beyond the point of excess. And, as any anthropologist will tell you, gifts also bind us into long-lasting covenants of reciprocal exchange.[30] They force us to consider what we can give back in return. The gift lingers; if you've received a gift from someone, you won't accept another one until you've had a chance to give something back to them. In this sense, the logic of the gift is deeply ecological: it is about equilibrium, about balance. Indeed, it is how ecosystems maintain themselves.

All of this runs *exactly* against the logic of capitalism. Capitalism ultimately relies on a single, overarching principle: take more than you give back. We've seen this logic in action for 500 years, beginning with enclosure and colonisation. In order to accumulate surplus, you have to extract uncompensated value from nature and bodies, which must be objectified and rendered as 'external'.

So what would it mean to extend the principles of reciprocity beyond the individual interactions that we might have with plants and animals and ecosystems? What would it mean to govern a whole economic system by these rules? Interestingly, ecological economists are already taking steps in this direction. Remember, the key principle of ecological economics is to run the economy in steady-state: to extract no more than can be regenerated, and to waste no more than can be safely absorbed. The Achuar and Chewong would find a lot to resonate with here.

How can we know what those thresholds are? That's where ecologists come in. Ecology is a unique branch of science, in that it seeks not only to understand the *parts* of a system, but how those parts relate to one another in a broader whole. Ecologists are adept at understanding and even managing ecosystem health. They are in some crucial respects like shamans. Drawing on insights from ecologists, whether their expertise comes from university training or from longstanding engagement with the land, we can determine how many trees can be felled, how many fish can be fished, and how much ore can be mined without tipping ecosystems out of balance, and we can set limits and quotas accordingly.

Better yet, we can switch to methods that don't just minimise harm, but actively *regenerate* ecosystems. This is where the reciprocity part comes in; and it's where things get particularly exciting. Take farming, for instance. Modern industrial farms are built as vast monocultures, with a single crop stretching from horizon to horizon, doused in chemical pesticides and herbicides designed to exterminate all other forms of life. If you've ever seen aerial photographs of the American Midwest, you know what this looks like: under capitalist agriculture, the land is reorganised according to a totalitarian logic with a single goal in mind: to maximise short-term extraction. This approach has turned rich topsoils into dust, releasing huge plumes of CO_2 from the earth in the process. It's caused insect and bird populations to collapse, while chemical run-off has killed whole freshwater ecosystems.

Fortunately, there's another way. Intrepid farmers around the world, from Virginia to Syria, are experimenting with more holistic methods called regenerative agroecology. They're planting multiple crop species together to build resilient ecosystems, while using compost, organic fertilisers and crop rotation to restore life and fertility to the soils. In areas where these methods have been used, crop yields have improved, earthworms have returned, insect populations have recovered and bird species have rebounded.[31] And perhaps best of all, as dead soils recover they are sequestering enormous quantities of CO_2 out of the atmosphere. In fact, scientists believe that if we're going to have any chance of averting climate breakdown, we'll need to roll out regenerative methods across most of the world's farmland and pasture. It's more effective by far than any artificial carbon-capture technology.

This is what reciprocity looks like in action. When you give back as much as you receive, it has a multiplier effect on ecosystem health. It revives life. And it's not just in agriculture that this is happening. Regenerative approaches are being developed in forestry and fishing as well, and in many cases people are drawing on techniques that have long been used by Indigenous communities and small farmers in the global South.

Large agribusinesses have been slow to adopt these methods, however – despite the fact that they have been shown to improve the quality of crops and the long-term fertility of the soils. Why? Because it requires time and labour. It requires an intimate knowledge of the local ecosystem. It requires understanding the traits and behaviours of dozens of species, and how they interact with each other. It requires care. When you treat a farm like an ecosystem instead of a factory, you begin a relationship with the land that is inimical to the short-term extractivist logic of agribusiness.

*

Some communities are taking these principles even further. Instead of just encouraging reciprocity with ecosystems, they are giving nature the rights of *legal personhood*. If this sounds wild, take a minute to remember that we already give personhood status to certain non-human entities: namely, corporations. This is a twisted view of personhood that privileges accumulation over life itself. We can flip this logic around. Instead of giving personhood to Exxon and Facebook, we can give legal recognition to living beings. Why not redwoods? Why not rivers? Why not whole watersheds?

Over the past few years, a series of extraordinary court decisions in New Zealand has caused an international stir. In 2017, the Whanganui River – the country's third longest river, which the Maori people have long considered to be sacred – was declared a legal person. It is now recognised as 'an indivisible and living whole from the mountains to the sea', incorporating both its physical and metaphysical elements. The Maori have been fighting for this since 1870. In the words of Gerrard Albert, the lead negotiator, 'We consider the river an ancestor and always have.' And it's not just the river. In the same year, courts gave similar legal standing to Mount Taranaki, which towers over the island's west coast. A few years prior, the Te Urewera national park was made a legal entity, no longer to be owned by the government as state property, but rather to be *owned by itself*.

Following the New Zealand decision, the Ganges and Yamuna rivers in India were given legal rights: 'all the corresponding rights, duties and liabilities of a living person.' In Colombia, the Supreme Court granted legal rights to the Amazon River. Going forward, any acts that harm these rivers can technically be prosecuted in much the same way that we might prosecute harms perpetrated against humans.

Some countries have gone further still. Ecuador's 2008 constitution establishes the rights of nature itself 'to exist, persist, maintain and regenerate its vital cycles'. Two years later, Bolivia passed the Law of the Rights of Mother Earth, recognising that 'Mother Earth is the dynamic living system formed by the indivisible community of all life systems and living beings who are interrelated, interdependent and complementary, which share a common destiny'. While some worry that these rights may turn out to be more rhetorical than real, there is nonetheless a lot of potential here, and they have already been successfully invoked in some cases to stop big industrial projects that might harm rivers and watersheds.

Can we extend this approach even more broadly, to encompass the whole planet? Some people think so. There is a movement of Indigenous communities and their allies to get a Universal Declaration of the Rights of Mother Earth formally adopted by the UN General Assembly. The draft declaration says that the Earth should have 'the right to life and to exist, the right to be respected, the right to regenerate its bio-capacity and to continue its vital cycles and processes'. At the same time, a growing movement of scientists is calling for a framework to protect major planetary processes like the carbon cycle, the nitrogen cycle, ocean currents, forests, the ozone layer and so on in order to maintain the conditions for life. And because all of these processes traverse human-created borders, protecting them requires co-operation beyond the nation-state.

Less is more

All of this represents the beginning of a profound shift in consciousness. There's something about the ecological crisis that seems to be opening us to new ways of thinking (or rather beckoning us to older ways of thinking) about our relationship with the more-than-human world. This takes us straight to the core of the problem. It gestures towards how we might begin to heal the rift from which this crisis has ultimately sprung. It empowers us to imagine a richer, more fertile future: a future free from the old dogmas of capitalism and rooted instead in reciprocity with the living world.

The ecological crisis requires a radical policy response. We need high-income countries to scale down excess energy and material use; we need a rapid transition to renewables; and we need to shift to a post-capitalist economy that's focused on human well-being and ecological stability rather than on perpetual growth. But we also need more than this – we need a new way of thinking about our relationship with the living world. How can we possibly bring all of these together?

When I set out to write this book, I worried about using degrowth as a central frame. It is only a first step, after all. But as I think about the journey we've been on, I wonder if it is also more than that. Degrowth provides a way for us to approach this challenge. It stands for de-colonisation, of both lands and peoples and even our minds. It stands for the de-enclosure of commons, the de-commodification of public goods, and the de-intensification of work and life. It stands for the de-thingification of humans and nature, and the de-escalation of ecological crisis. Degrowth begins as a process of taking less. But in the end it opens up whole vistas of possibility. It moves us from scarcity to abundance, from extraction to regeneration, from dominion to reciprocity, and from loneliness and separation to connection with a world that's fizzing with life.

Ultimately, what we call 'the economy' is our material relationship with each other and with the rest of the living world. We must ask ourselves: what do we want that relationship to be like? Do we want it to be about domination and extraction? Or do we want it to be about reciprocity and care?

*

There is a tree that stands outside the window of the room in London where I write. It's an enormous chestnut that whirls confidently out of the earth and casts its generous branches nearly five storeys high. The species has been around for some 80 million years, having somehow survived the last mass extinction event. This particular tree is 500 years old, and one of the last remnants of an ancient forest that has long since been destroyed. It has stood as witness to the whole story that I have described in these pages. It was there even before the enclosure movement began, when the land from which its roots draw sustenance was still a commons unencumbered by title or deed. It was there as the early colonial invasions set sail. It has watched, season after season, as industrial emissions have poured into the sky, as the temperatures have risen, and as the insects and birds that live amongst its leaves have slowly disappeared.

I often wonder what this quiet giant will witness in the decades and centuries ahead, during our lifetime, and the lifetimes of the generations that will follow. How will the rest of the story unfold? It is within our power to write a different future, if we can summon the courage to do so. We have everything to lose, and a world to gain.

Acknowledgements

It is said that the Buddha told this story as a warning. A couple were travelling across the desert with their only child. Their food supply ran low, and they grew hungry. But driven by an insatiable ambition for their destination, they refused to change course. As if in a trance, they decided to kill and eat their child to sustain them. When they arrived at last on the other side, when the destination lost its allure and the trance lost its grip, they were utterly hollowed with grief and regret.

What are we doing here? Where are we going? What's it all for? What is the end, as it were, of human existence? Growthism prevents us from stopping to think about these questions. It prevents us from reflecting on what we actually want our society to achieve. Indeed, the pursuit of growth comes to stand in for thought itself. We are in a trance. We slog on, mindlessly, unaware of what we're doing, unaware of what's happening around us, unaware of what we are sacrificing . . . *who* we are sacrificing.[1]

Degrowth is an idea that shakes us out of the trance. 'Sit, be still, and listen,' Rumi wrote in one of his poems: 'for you are drunk, and we are at the edge of the roof.'

This is not about living a life of voluntary misery or imposing harsh limits on human potential. On the contrary, it is exactly the opposite. It is about flourishing, and about reaching a higher level of consciousness about what we're doing here and why.

But the trance is powerful. To escape it requires escaping the ruts grooved into our minds, the assumptions baked into our culture, the ideologies that shape our politics. That is no easy task. It requires courage and discipline. For me, it has been a long journey, and I still have many miles to go. At every step along the way I have relied on the grace of fellow travelers who have pulled me out of ruts and opened me to new ways of seeing the world.

I've benefitted immensely from personal conversations – and in some cases collaborations – with Giorgos Kallis, Kate Raworth, Daniel O'Neill, Julia Steinberger, John Bellamy Foster, Ian Gough, Ajay Chaudhary, Glen Peters, Ewan McGaughey, Asad Rehman, Bev Skeggs, David Graeber, Sam Bliss, Riccardo Mastini, Jason Hirsch, Federico de Maria, Peter Victor, Ann Pettifor, Lorenzo Fioramonti, Peter Lipman, Joan Martinez-Alier, Martin Kirk, Alnoor Ladha, Huzaifa Zoomkawala, Patrick Bond, Rupert Read, Fred Damon, Wende Marshall, Federico Cruz, The Rules team, my editors at the *Guardian*, *Foreign Policy*, Al Jazeera and other outlets, where I first worked out many of the ideas that appear in this book, and of course my agent Zoe Ross, and Tom Avery, my editor at Penguin, who were willing to give this idea a platform.

I've also learned from and been inspired by the writings of many others: Silvia Federici, Jason Moore, Raj Patel, Andreas Malm, Naomi Klein, Kevin Anderson, Tim Jackson, Juliet Schor,

Vandana Shiva, Arturo Escobar, George Monbiot, Herman Daly, Kate Aronoff, Robert Macfarlane, Abdullah Öcalan, Ariel Salleh, David Wallace-Wells, Nnimmo Bassey, Robin Wall Kimmerer, Timothy Morton, Daniel Quinn, Carolyn Merchant, Vijay Prashad, David Harvey, Maria Mies, Gustavo Esteva, André Gorz, Serge Latouche, Bill McKibben, Jack D. Forbes, Philippe Descola, David Abrams, Kofi Klu, Bruno Latour, Suzanne Simard, Murray Bookchin, and Ursula Le Guin. Their works have been signposts along the way.

But this list only barely scratches the surface. And I cannot leave out the towering figures whose words – and lives – I find myself returning to over and over again, for grounding and direction: Aimé Césaire, Frantz Fanon, Thomas Sankara, Berta Cáceres, Mahatma Gandhi, Patrice Lumumba, Samir Amin. They guide me as the ancestors.

I am also grateful to the students I've engaged with while teaching: at the London School of Economics, at the Autonomous University of Barcelona, at Schumacher College, at Goldsmiths and elsewhere. I've encountered more than a few classrooms that have expanded my horizons and given me new ways to think and speak.

I finished writing this book during the coronavirus lockdown in London. I will always remember it as a strange and eerie time. We all suddenly realized what parts of the economy really matter – and whose jobs we depend on most. For me, this was inescapably clear. My partner, Guddi, is a doctor in the NHS. In those early weeks I would watch her walk out the door each morning on her way to what amounted to a warzone, hoping that my eyes didn't betray the fear I felt for her. And when she came home each evening, exhausted from work vastly more important than my own, she would still ask to read my drafts.

We used our allotted exercise time for walks, during which she would help me work through ideas and sharpen arguments and find narrative arcs, while we watched grey winter give way to the tender leaves and blossoms of spring. This book – and especially its final chapter – represents a shared intellectual journey. I am endlessly grateful for her wisdom, insight, companionship, and her unflagging ability to see through every ruse that our culture has going. She sharpens me every day.

Early in 2012, Guddi and I attended a public lecture by Paul Krugman at the LSE. It was during the Great Recession, and Krugman argued that the United States needed a massive government stimulus to get growth going again. As we walked home, Guddi wondered aloud whether the US, one of the richest nations in the world, really required more GDP, when so many nations do so much better on all the indicators that really matter, with much less. Do high-income economies really need to keep growing, forever? Toward what end? I responded with all the usual mantras – how growth is essential to a healthy economy and all that. But the question unsettled me. I still remember, during the quietness that followed, realising that I was just repeating things that had been told to me, without actually thinking for myself. That conversation was the beginning of the eight-year journey that led to this book.

There is nothing more powerful than a question.

Endnotes

Introduction: Welcome to the Anthropocene

1 Damian Carrington, 'Warning of 'ecological Armageddon' after dramatic plunge in insect numbers,' *Guardian*, 2017.

2 Patrick Barkham, 'Europe faces 'biodiversity oblivion' after collapse in French birds, experts warn,' *Guardian*, 2018.

3 Roel Van Klink et al., 'Meta-analysis reveals declines in terrestrial but increases in freshwater insect abundances,' *Science* 368(6489), 2020, pp. 417–420. The study indicated that freshwater insect abundances were increasing, although these claims were questioned in the same journal: Marion Desquilbet et al., *Science* 370(6523), 2020.

4 IPBES, *Global Assessment Report on Biodiversity and Ecosystem Services*, 2019. Some estimates indicate that as much as 40% of insect species may be at risk of extinction. Responding to these claims, Josef Settele, co-chair of the IPBES report, said: 'Forty per cent might be too high, and 10% in our global assessment is too low, but this is the range.' See Ajit Niranjan, 'Insects are dying and nobody knows how fast,' *DW*, 2020. The dearth of robust historical data makes longitudinal assessment difficult. And biomass trends may fluctuate: one study found that in Britain moth biomass increased from 1967 to 1982, and has declined steadily since then. See Callum Macgregor et al., 'Moth biomass increases and decreases over 50 years in Britain,' *Nature Ecology & Evolution* 3, 2019, pp. 1645–1649.

5 Rachel Kehoe et al., 'Cascading extinctions as a hidden driver of insect decline,' *Ecological Entomology* 46(4), 2021, pp. 743–756.

6 Pedro Cardoso et al., 'Scientists' warning to humanity on insect extinctions,' *Biological Conservation* 242, 2020.

7 David Wagner et al., 'Insect decline in the Anthropocene: Death by a thousand cuts,' *Proceedings of the National Academy of Sciences* 118(2), 2021.

8 IPCC, *Special Report: Climate Change and Land,* 2018.

9 Robert Blakemore, 'Critical decline of earthworms from organic origins under intensive, humic SOM-depleting agriculture,' *Soil Systems* 2(2), 2018.

10 FAO, *The State of World Fisheries and Aquaculture* (UN Food and Agriculture Organization, 2020).

11 Ruth Thurstan et al., 'The effects of 118 years of industrial fishing on UK bottom trawl fisheries,' *Nature Communications* 1(1), 2010.

12 Daniel Pauly and Dirk Zeller, 'Catch reconstructions reveal that global marine fisheries catches are higher than reported and declining,' *Nature Communications* 7, 2016.

13 Jonathan Watts, 'Destruction of nature as dangerous as climate change, scientists warn,' *Guardian,* 2018. One might hope that we can replace declining catches with fish farms, but it's not quite so simple. Every ton of farmed fish needs as much as five tons of wild fish to be trawled up and ground into feed. And fish farms involve heavy use of medicines and chemical disinfectants, which are already a major source of marine pollution. See John Vidal, 'Salmon farming in crisis,' *Guardian,* 2017.

14 We've reached the point where we're dropping the equivalent of six atomic bombs' worth of heat into the sea every second. Damian Carrington, 'Global warming of oceans equivalent to an atomic bomb per second,' *Guardian,* 2019.

15 Marine life depends on temperature gradients that circulate nutrients from the seafloor to the surface. As oceans warm, those gradients are breaking down and nutrient cycles are stagnating.

16 Damian Carrington, 'Ocean acidification can cause mass extinctions, fossils reveal,' *Guardian,* 2019.

17 Malin Pinsky et al., 'Greater vulnerability to warming of marine versus terrestrial ectotherms,' *Nature* 569(7754), 2019, pp. 108–111.

18 Bärbel Hönisch et al., 'The geological record of ocean acidification,' *Science* 335(6072), 2012, pp. 1058–1063. Coral reefs support a quarter of all ocean life, including species that are crucial to human food systems. Half

a billion people rely on coral ecosystems for food. See David Wallace-Wells, 'The Uninhabitable Earth,' *New York* magazine, 2017.

19 Jurriaan De Vos et al., 'Estimating the normal background rate of species extinction,' *Conservation Biology* 29(2), 2015.

20 IPBES, *Global Assessment Report on Biodiversity and Ecosystem Services*, 2019.

21 Gerardo Ceballos et al., 'Biological annihilation via the ongoing sixth mass extinction signaled by vertebrate population losses and declines,' *Proceedings of the National Academy of Sciences* 114(30), 2017.

22 According to the European Academies' Science Advisory Council.

23 IPCC, *Special Report: Global Warming of 1.5°C*, 2018.

24 NASA, 'NASA study finds carbon emissions could dramatically increase risk of US megadroughts,' 2015.

25 Chuan Zhao et al., 'Temperature increase reduces global yields of major crops in four independent estimates,' *Proceedings of the National Academy of Sciences* 114(35), 2017.

26 Deepak Ray, 'Climate change is affecting crop yields and reducing global food supplies,' *Conversation*, 2019.

27 Ferris Jabr, 'The Earth is just as alive as you are,' *New York Times*, 2019.

28 Robert DeConto and David Pollard, 'Contribution of Antarctica to past and future sea-level rise,' *Nature* 531(7596), 2016, pp. 591–597.

29 Will Steffen et al., 'Trajectories of the Earth System in the Anthropocene,' *Proceedings of the National Academy of Sciences* 115(33), 2018, p. 8252–8259.

30 Timothy Morton, *Being Ecological* (Penguin, 2018).

31 Growth is not the *only* distinguishing feature of capitalism, of course. Proletarian wage labour and 'private property' (i.e., exclusive control over the means of production) are also key features. But when it comes to the question of capital's relationship with ecology, growth dependency is a major problem. Of course, a number of socialist regimes in the twentieth century were also productivist. The USSR, for instance, was famously growth-obsessed. In this respect it had a kind of state capitalist character (being organised around surplus and reinvestment for the sake of expansion), which is one of the reasons that it does not offer a meaningful alternative to our present crisis.

32 Mathias Binswanger, 'The growth imperative revisited: a rejoinder to Gilányi and Johnson,' *Journal of Post Keynesian Economics* 37(4), 2015, pp. 648–660.

33 Johan Rockström et al., 'Planetary boundaries: exploring the safe operating space for humanity,' *Ecology and Society* 14(2), 2009; Will Steffen et al., 'Planetary boundaries: Guiding human development on a changing planet,' *Science* 347(6223), 2015.

34 To see which countries are overshooting planetary boundaries, see goodlife.leeds.ac.uk/countries.

35 See www.calculator.climateequityreference.org.

36 For relevant evidence, see Jason Hickel, Paul Brockway, Giorgos Kallis, Lorenz Keyßer, Manfred Lenzen, Aljoša Slameršak, Julia Steinberger, and Diana Ürge-Vorsatz, 'Urgent need for post-growth climate mitigation scenarios.' *Nature Energy* 6(8), 2021. To have a 66% chance of staying under 1.5°C, global emissions must fall by 10% per year beginning in 2020. If the global economy grows at 2.6% per year (as PwC predicts), this requires decarbonisation of 14% per year. This is nearly nine times faster than the business as usual rate of decarbonisation (1.6% per year), and more than three times faster than the maximum rate assumed in best-case scenario models (4% per year). In other words, it is out of scope. To have a 50% chance of staying under 1.5°C emissions must fall by 7.3% per year, with decarbonisation of 10.7% per year, which is also out of scope. To have a 66% chance of staying under 2°C (as per the Paris Agreement) emissions must fall by 4.1% per year, with decarbonisation of 7% per year: again, out of scope (however, it may be feasible to achieve *if* the economy does not grow). These are global figures. For high-income nations it is much more difficult: to be consistent with the Paris Agreement target for 2°C, and to uphold the principle of equity, they must reduce emissions by 12% per year. Even in a no-growth scenario this is impossible; it requires degrowth. See Jason Hickel and Giorgos Kallis, 'Is green growth possible?' *New Political Economy*, 2019 (note that the figures I have cited here are updated since the publication of the article).

37 When I say 'more growth means more energy demand', I mean compared to the baseline of what the economy would otherwise require under any given mix of energy sources.

38 Hickel and Kallis, 'Is green growth possible?' In addition, a review of 835 empirical studies finds that decoupling is not adequate to achieve climate goals; it requires what the authors themselves refer to as 'degrowth' scenarios: Helmut Haberl et al., 'A systematic review of the evidence on decoupling of GDP, resource use and GHG emissions: part II: synthesizing the insights,' *Environmental Research Letters*, 2020. Another review of 179 studies finds 'no evidence of economy-wide, national or international absolute resource decoupling, and no evidence of the kind of decoupling needed for ecological sustainability': T. Vadén et al., 'Decoupling for ecological sustainability: A categorisation and review of research literature,' *Environmental Science and Policy*, 2020.

39 'Survey of young Americans' attitudes toward politics and public service,' Harvard University Institute of Politics, 2016.

40 *Edelman Trust Barometer*, 2020.

41 According to the same 2015 YouGov poll mentioned earlier in the paragraph.

42 Yale Climate Opinion Maps, Yale Program on Climate Change Communication.

43 Stefan Drews et al., 'Challenges in assessing public opinion on economic growth versus environment: considering European and US data,' *Ecological Economics* 146, 2018, pp. 265–272.

44 *The New Consumer and the Sharing Economy*, Havas, 2015.

45 'The EU needs a stability and well-being pact, not more growth,' *Guardian*, 2018.

46 William Ripple et al., 'World scientists warn of a climate emergency,' *BioScience*, 2019.

47 World Inequality Database.

48 Lorenz Keyßer and Manfred Lenzen, '1.5 C degrowth scenarios suggest the need for new mitigation pathways,' *Nature Communications* 12(1), 2021; Kai Kuhnhenn et al., *A societal transformation scenario for staying below 1.5 C* (Heinrich-Böll-Stiftung, 2020). The lead scenario in the IPCC's 2018 report relies on declining material and energy throughput. This is the only scenario that does not rely on speculative negative-emissions technologies. The underlying paper is Arnulf Grubler et al.,

'A low energy demand scenario for meeting the 1.5 C target and sustainable development goals without negative emission technologies,' *Nature Energy* 3(6), 2018, pp. 515–527. For my take on this scenario, see Hickel and Kallis, 'Is green growth possible?'

49 See: Serge Latouche, *Farewell to Growth* (Polity, 2009); Giorgos Kallis, Christian Kerschner and Joan Martinez-Alier, 'The economics of degrowth,' *Ecological Economics* 84, 2012, pp. 172–180; Giacomo D'Alisa et al., eds., *Degrowth: A Vocabulary for a New Era* (Routledge, 2014); Giorgos Kallis, *Degrowth* (Agenda Publishing, 2018); Jason Hickel, 'What does degrowth mean? A few points of clarification', *Globalizations*, 2020.

50 Joel Millward-Hopkins et al. 'Providing decent living with minimum energy: A global scenario', *Global Environmental Change* 65, 2020; Michael Lettenmeier et al. 'Eight tons of material footprint—suggestion for a resource cap for household consumption in Finland', *Resources* 3(3), 2014.

51 For a history and overview of degrowth, see Kallis, *Degrowth*; for global South perspectives see Arturo Escobar, 'Degrowth, postdevelopment, and transitions: a preliminary conversation,' *Sustainability Science*, 2015.

52 For this framing I am indebted to Timothy Morton, *Ecology Without Nature* (Harvard University Press, 2007).

One: Capitalism – A Creation Story

1 Jason Moore, *Capitalism in the Web of Life* (Verso, 2015).

2 I draw this from Braudel. See also David Graeber, *Debt: The First 5,000 Years* (Penguin UK, 2012), pp. 271–282.

3 I first learned about this history from Silvia Federici, *Caliban and the Witch* (Autonomedia, 2004); I draw on her work for much of this chapter. I am also grateful for insights from Jason Hirsch, and his book *Wildflower Counter-Power* (Triarchy Press, 2020).

4 Samuel Kline Cohn, *Lust for Liberty: The Politics of Social Revolt in Medieval Europe, 1200–1425* (Harvard University Press, 2009).

5 Federici, *Caliban and the Witch*, p. 46.

6   James E. Thorold Rogers, *Six Centuries of Work and Wages: The History of English Labour* (London, 1894), pp. 326ff; P. Boissonnade, *Life and Work in Medieval Europe* (New York: Alfred A. Knopf, 1927), pp. 316–20.

7   Fernand Braudel, *Capitalism and Material Life, 1400-1800* (New York: Harper and Row, 1967), pp. 128ff; Karl Marx, *Capital* Vol. 1.

8   Carolyn Merchant, *The Death of Nature: Women, Ecology, and the Scientific Revolution* (1981).

9   Christopher Dyer, 'A redistribution of income in 15th century England,' *Past and Present* 39, 1968, p. 33.

10   John Hatcher, 'England in the aftermath of the Black Death,' *Past and Present* 144, 1994, p. 17.

11   This is Federici's term.

12   Enclosure of manorial common land was initially authorised by the Statute of Merton (1235) and the Statute of Westminster (1285), only shortly after rights to commons were enshrined in the Charter of the Forest (1217). For more, see Guy Standing, *Plunder of the Commons* (Penguin, 2019).

13   Henry Phelps Brown and Sheila V. Hopkins, *A Perspective of Wages and Prices* (Routledge, 2013).

14   Edward Wrigley and Roger Schofield, *The Population History of England 1541-1871* (Cambridge University Press, 1989).

15   I draw this observation from Mark Cohen, *Health and the Rise of Civilisation* (Yale University Press, 1989).

16   Simon Szreter, 'The population health approach in historical perspective,' *American Journal of Public Health* 93(3), 2003, pp. 421–431; Simon Szreter and Graham Mooney, 'Urbanization, mortality, and the standard of living debate: new estimates of the expectation of life at birth in nineteenth-century British cities,' *Economic History Review* 51(1), 1998, pp. 84–112.

17   Timothy Walton, *The Spanish Treasure Fleets* (Florida: Pineapple Press, 1994); Kenneth Pomeranz, *The Great Divergence: China, Europe, and the Making of the Modern World Economy* (Princeton University Press, 2009). For more on this history, and relevant sources, see *The Divide*.

18   Pomeranz, Chapter 6 in *The Great Divergence*; Sven Beckert, *Empire of Cotton: A Global History* (Vintage, 2015).

19 Andrés Reséndez, *The Other Slavery: The Uncovered Story of Indian Enslavement in America* (Houghton Mifflin Harcourt, 2016).

20 These figures come from a 1993 article in *Harper's* magazine. The minimum wage is calculated at the 1993 rate, interest through 1993, and the results are expressed in 1993 dollars; an updated figure would be much higher than this.

21 Utsa Patnaik, *Agrarian and Other Histories* (Tulik Books, 2018); Jason Hickel, 'How Britain stole $45 trillion from India,' *Al Jazeera*, 2018; Gurminder Bhambra, '"Our Island Story": The Dangerous Politics of Belonging in Austere Times,' in *Austere Histories in European Societies* (Routledge, 2017).

22 B.R. Tomlinson, 'Economics: The Periphery,' In *The Oxford History of the British Empire* (1990), p. 69.

23 Ellen Meiksins Wood, *The Origin of Capitalism: A Longer View* (Verso, 2003).

24 Karl Polanyi, *The Great Transformation* (Boston: Beacon Press, 1944).

25 John Locke, *The Second Treatise of Government*, 1689.

26 For more on this history of scarcity, see Nicholas Xenos, *Scarcity and Modernity* (Routledge, 2017).

27 I derive these quotes from Michael Perelman, *The Invention of Capitalism: Classical Political Economy and the Secret History of Primitive Accumulation* (Duke University Press, 2000).

28 Mike Davis, *Late Victorian Holocausts: El Niño Famines and the Making of the Third World* (Verso Books, 2002).

29 Maitland explored this paradox in a book titled *Inquiry into the Nature and Origin of Public Wealth and into the Means and Causes of its Increase*. For more on this, see John Bellamy Foster, Brett Clark and Richard York, *The Ecological Rift: Capitalism's War on the Earth* (NYU Press, 2011).

30 This history is charted in Merchant, *Death of Nature*.

31 Stephen Gaukroger, *The Emergence of a Scientific Culture: Science and the Shaping of Modernity 1210–1685* (Clarendon Press, 2008).

32 Brian Easle, *Witch-Hunting, Magic and the New Philosophy* (The Harvester Press, 1980), cited in Federici p. 149.

33 Merchant, *Death of Nature*, p. 3.

34 Gaukroger, p. 325.

35 Immanuel Kant, *Lecture on Ethics*, 1779.

36 Juliet Schor, *The Overworked American: The Unexpected Decline of Leisure* (Basic Books, 2008).

37 E.P. Thompson, *Customs in Common: Studies in Traditional Popular Culture* (New Press/ORIM, 2015).

38 This language was used in the Vagabonds Act of 1536.

39 William Harrison, *Description of Elizabethan England*, 1577.

40 See Max Weber, *The Protestant Ethic and the Spirit of Capitalism* (1930).

41 See Raj Patel and Jason W. Moore, *A History of the World in Seven Cheap Things: A Guide to Capitalism, Nature, and the Future of the Planet* (University of California Press, 2017).

42 Federici explores this issue at length in *Caliban and the Witch*. See also Maria Mies, *Patriarchy and Accumulation on a World Scale* (London: Zed, 1986).

43 Aimé Césaire, *Discourse on Colonialism*, 1955.

44 Mario Blaser, 'Political ontology: Cultural studies without 'cultures'?' *Cultural Studies* 23(5–6), 2009, pp. 873–896.

45 Ngũgĩ wa Thiong'o, *Decolonising the Mind: The Politics of Language in African Culture* (London: James Currey, 1986).

46 I draw this insight from Timothy Morton, *Being Ecological* (Penguin, 2018).

47 I draw this insight from a 2002 speech by Daniel Quinn titled 'A New Renaissance'.

Two: Rise of the Juggernaut

1 Jason W. Moore, 'The Capitalocene Part II: accumulation by appropriation and the centrality of unpaid work/energy,' *Journal of Peasant Studies* 45(2), 2018, pp. 237–279.

2 I draw the concepts of 'use-value' and 'exchange-value' – and the general formula of capital accumulation – from Marx's *Capital*. For more on the relationship between capital and ecological breakdown, see Foster and Clark, 'The planetary emergency,' *Monthly Review*, 2012.

3 The giant vampire squid is Matt Taibbi's metaphor.

4   David Harvey, *A Brief History of Neoliberalism* (Oxford University Press, 2007).

5   Matthias Schmelzer, *The Hegemony of Growth: The OECD and the Making of the Economic Growth Paradigm* (Cambridge University Press, 2016).

6   David Harvey, *A Brief History of Neoliberalism* (Oxford, 2005).

7   For more on the story of post-colonial developmentalist policy in the South, and its reversal beginning in the 1980s, see Jason Hickel, *The Divide* (London: Penguin Random House, 2018), Chapters 4 and 5.

8   Hickel, *The Divide*, Chapter 5.

9   Harvey, *A Brief History of Neoliberalism*.

10 Jason Hickel, 'Global inequality: do we really live in a one-hump world?' *Global Policy*, 2019.

11 For more on how this works, see Jason Hickel, 'The new shock doctrine: 'Doing business' with the World Bank,' *Al Jazeera*, 2014.

12 Tim Jackson and Peter Victor, 'Productivity and work in the 'green economy': some theoretical reflections and empirical tests,' *Environmental Innovation and Societal Transitions* 1(1), 2011, pp. 101–108.

13 For figures from 1900 to 1970 I rely on F. Krausmann et al., 'Growth in global materials use, GDP and population during the 20th century,' *Ecological Economics*, 68(10), 2009, pp. 2696–2705. For figures from 1970 to 2017 I rely on materialflows.net. For figures to 2020 I rely on UN International Resource Panel projections.

14 Stefan Bringezu, 'Possible target corridor for sustainable use of global material resources,' *Resources* 4(1), 2015, pp. 25–54. Bringezu suggests a safe target range of 25–50 billion tons. Of course, it's difficult to define an aggregate limit for material footprint, since different materials have different kinds of impacts, and the impact of extraction varies according to the technologies used to manage it. In addition, one might argue that boundaries for some kinds of extraction should be defined regionally, not globally. Nonetheless, there is a consensus around 50 billion tons as a reasonable maximum global threshold figure.

15 International Resource Panel, *Global Resources Outlook* (UN Environment Programme, 2019).

16 See Z. J. Steinmann et al., 'Resource footprints are good proxies of environmental damage,' *Environmental Science & Technology* 51(11), 2017.

Resource footprints (materials, energy, land and water) account for more than 90% of variation in environmental damage footprints, and account for 90% of damage to biodiversity. In aggregate, the total mass of material use is coupled to ecological impact, with a correlation factor of 0.73. See E. Voet et al., 'Dematerialisation: not just a matter of weight,' *Journal of Industrial Ecology*, 8(4), 2004, pp. 121–137.

17 The relationship between GDP and energy is not one-to-one; efficiency improvements have led to a steady rate of relative decoupling over time. Nonetheless, the relationship is strongly positive (i.e., every additional unit of GDP entails more energy use).

18 Recent research, however, has cast doubt on the longstanding assumption that gas is less emissions intensive than oil: Benjamin Hmiel et al., 'Preindustrial 14 CH4 indicates greater anthropogenic fossil CH4 emissions,' *Nature* 578 (7795), 2020, pp. 409–412.

19 'Global primary energy consumption,' Our World in Data, 2018.

20 FAO, *Current Worldwide Annual Meat Consumption Per Capita, Livestock and Fish Primary Equivalent* (UN Food and Agriculture Organization, 2013).

21 'Global consumption of plastic materials by region,' *Plastics Insight*, 2016.

22 The figures here are for material footprint, which includes the raw material impacts of imported products. For a discussion of the per-capita boundary, see Bringezu, 'Possible target corridor for sustainable use of global material resources.'

23 I use 8 tons per capita as the sustainability threshold here, which is suggested as the 2030 target by Giljum Dittrich et al.

24 For this paragraph I draw on ideas and language from Kate Raworth, personal correspondence.

25 Christian Dorninger et al., 'Global patterns of ecologically unequal exchange: implications for sustainability in the 21st century,' *Ecological Economics*, 2020.

26 Jason Hickel, 'Quantifying national responsibility for climate breakdown: An equality-based attribution approach to carbon dioxide emissions in excess of the planetary boundary', *Lancet Planetary Health*, 2020. Here I depict the results as the sum of national overshoots within each region.

27 These results are based on my own calculations, in Hickel, 'Who is responsible for climate breakdown?' I use territorial emissions from 1850 to 1970, and consumption-based emissions from 1970 to 2015.

28 Climate Vulnerability Monitor (DARA, 2012)

29 'Climate change and poverty,' Human Rights Council, 2019.

30 Tom Wilson, 'Climate change in Somaliland – 'you can touch it,' *Financial Times*, 2018.

31 Rockström et al., 'Planetary boundaries'; Steffen et al., 'Planetary boundaries.'

32 See Giorgos Kallis, *Limits: Why Malthus was Wrong and Why Environmentalists Should Care* (Stanford University Press, 2019). This new way of thinking about limits is reflected to some extent in the Paris Agreement. Recognising the reality of planetary boundaries, nations have pledged to limit global warming to 1.5° C – at least on paper. We can expand this approach by pushing for similar commitments on all of the other planetary boundaries.

Three: Will Technology Save Us?

1 Leo Hickman, 'The history of BECCS,' *Carbon Brief*, 2016.

2 Glen Peters, 'Does the carbon budget mean the end of fossil fuels?' *Climate News*, 2017.

3 There may also be issues finding enough storage capacity for all the CO_2 that we would pull out of the atmosphere. And it could be vulnerable to leakage, in case of earthquakes etc. H. De Coninck and S.M. Benson, 'Carbon dioxide capture and storage: issues and prospects,' *Annual Review of Environment and Resources*, 39, 2014, pp. 243–270.

4 Sabine Fuss et al., 'Betting on negative emissions,' *Nature Climate Change* 4(10), 2014, pp. 850–853.

5 Pete Smith et al., 'Biophysical and economic limits to negative CO_2 emissions,' *Nature Climate Change* 6(1), 2016, pp. 42–50.

6 Kevin Anderson and Glen Peters, 'The trouble with negative emissions,' *Science* 354(6309), 2016, pp. 182–183.

7 Vera Heck, 'Biomass-based negative emissions difficult to reconcile with planetary boundaries,' *Nature Climate Change* 8(2), 2018, pp. 151–155.

8 Pete Smith et al., 'Biophysical and economic limits to negative CO2 emissions,' *Nature Climate Change* 6(1), 2016, pp. 42–50.

9 'Six problems with BECCS,' FERN briefing, 2018.

10 Henry Shue, 'Climate dreaming: negative emissions, risk transfer, and irreversibility,' *Journal of Human Rights and the Environment* 8(2), 2017, pp. 203–216.

11 Hickman, 'The history of BECCS.'

12 Daisy Dunne, 'Geo-engineering carries 'large risks' for the natural world, studies show,' *Carbon Brief*, 2018.

13 See the Climate Equity Reference Calculator at calculator.climateequityreference.org.

14 PwC forecasts that global GDP will grow by an average of 2.6% per year to 2050 (reaching a total of 2.15 times larger). Given the existing relationship between GDP and energy, this means energy demand will increase by a factor of 1.83 by 2050. Of course, renewable energy is more efficient than fossil fuels, to the point where transitioning to renewables by 2050 could lead to no increase in total energy use, despite business-as-usual growth, but it would still be 1.83 times higher than it would otherwise be without growth (under any given energy mix).

15 These decarbonisation figures assume a 66% chance of staying under the target threshold, and average annual global GDP growth of 2.6% per year. The maximum decarbonisation rate assumed in best-case scenario models is 4% per year. For a review of relevant literature, see Hickel and Kallis, 'Is green growth possible?'

16 Christian Holz et al., 'Ratcheting ambition to limit warming to 1.5 C: trade-offs between emission reductions and carbon dioxide removal,' *Environmental Research Letters* 13(6), 2018.

17 The IPCC's 2018 report has only one scenario for staying under 1.5°C without using BECCS. It works by relying on a significant reduction of energy and material use. The underlying paper is Grubler et al., 'A low energy demand scenario for meeting the 1.5°C target.' See Hickel and Kallis, 'Is green growth possible?' for a discussion.

18 World Bank, *The Growing Role of Minerals and Metals for a Low-Carbon Future*, 2017.

19 'Leading scientists set out resource challenge of meeting net zero emissions in the UK by 2050,' Natural History Museum, 2019.

20 According to data from www.miningdataonline.com.

21 Amit Katwala, 'The spiralling environmental cost of our lithium battery addiction,' *WIRED*, 2018.

22 Jonathan Watts, 'Environmental activist murders double in 15 years,' *Guardian*, 2019.

23 Derek Abbott, 'Limits to growth: can nuclear power supply the world's needs?' *Bulletin of the Atomic Scientists* 68(5), 2012, p. 23–32.

24 Both of these quotes come from: Kate Aronoff, 'Inside geo-engineers' risky plan to block out the sun,' *In These Times*, 2018.

25 Trisos, C. H. et al., 'Potentially dangerous consequences for biodiversity of solar geo-engineering implementation and termination,' *Nature Ecology & Evolution*, 2018.

26 See Hickel and Kallis, 'Is green growth possible?'; Haberl et al., 'A systematic review of the evidence on decoupling'; and Vadén et al., 'Decoupling for ecological sustainability'.

27 International Resource Panel, *Decoupling 2* (UN Environment Programme, 2014).

28 Guiomar, Calvo et al., 'Decreasing ore grades in global metallic mining: A theoretical issue or a global reality?' *Resources* 5(4), 2016.

29 Monika Dittrich et al., *Green Economies Around the World?* (SERI, 2012).

30 Heinz Schandl et al., 'Decoupling global environmental pressure and economic growth: scenarios for energy use, materials use and carbon emissions,' *Journal of Cleaner Production* 132, 2016, pp. 45–56.

31 International Resource Panel, *Assessing Global Resource Use* (UN Environment Programme).

32 Tim Santarius, *Green Growth Unravelled: How Rebound Effects Baffle Sustainability Targets When the Economy Keeps Growing* (Heinrich Boll Stiftung, 2012).

33 Although some food can be made circular through composting and nutrient recovery.

34 W. Haas et al., 'How circular is the global economy? An assessment of material flows, waste production, and recycling in the European Union

and the world in 2005,' *Journal of Industrial Ecology*, 19(5), 2015, pp. 765–777.

35 *The Circularity Report* (PACE, 2015).

36 This idea was initially proposed by Herman Daly.

37 See the final chapter in Kallis, *Degrowth*.

38 Beth Stratford, 'The threat of rent extraction in a resource-constrained future,' *Ecological Economics* 169, 2020.

Four: Secrets of the Good Life

1 See Szreter, 'The population health approach in historical perspective'; Simon Szreter, 'Rapid economic growth and 'the four Ds' of disruption, deprivation, disease and death: public health lessons from nineteenth-century Britain for twenty-first-century China?' *Tropical Medicine & International Health* 4(2), pp. 146–152.

2 Simon Szreter, 'The importance of social intervention in Britain's mortality decline c. 1850–1914: A re-interpretation of the role of public health,' *Social history of medicine* 1(1), pp. 1–38.

3 Simon Szreter, 'Rethinking McKeown: The relationship between public health and social change,' *American Journal of Public Health* 92(5), pp. 722–725. Formally, public goods and commons are not the same thing (commons are collectively managed, whereas public goods are usually, although not always, centrally managed), but they are comparable here in the sense that they both constitute forms of collective provisioning.

4 David Cutler and Grant Miller, 'The role of public health improvements in health advances,' *Demography* 42(1), 2005.

5 Chhabi Ranabhat et al., 'The influence of universal health coverage on life expectancy at birth (LEAB) and healthy life expectancy (HALE): a multi-country cross-sectional study,' *Frontiers in Pharmacology* 9, 2018.

6 Wolfgang Lutz and Endale Kebede, 'Education and health: redrawing the Preston curve,' *Population and Development Review* 44(2), 2018.

7 Samuel Preston, 'The changing relation between mortality and level of economic development,' *Population Studies* 29(2), 1975.

8 UNDP, 'Training material for producing national human development reports,' UNDP Human Development Report Office, 2015. See also

UNDP, 'Understanding performance in human development,' Human Development Research Paper 42, 2010, pp. 28–32.

9 Szreter, 'The population health approach in historical perspective.'

10 See Shirley Cereseto and Howard Waitzkin, 'Economic development, political-economic system, and the physical quality of life,' *American Journal of Public Health* 76(6), 1986, and Amartya Sen, 'Public Action and the Quality of Life in Developing Countries,' *Oxford Bulletin of Economics and Statistics* 43(4), 1981.

11 Julia Steinberger and J. Timmons Roberts, 'From constraint to sufficiency: The decoupling of energy and carbon from human needs, 1975–2005,' *Ecological Economics* 70(2), 2010, pp. 425–433.

12 This data comes from the Centre on International Education Benchmarking.

13 Juliana Martínez Franzoni and Diego Sánchez-Ancochea, *The Quest for Universal Social Policy in the South: Actors, Ideas and Architectures* (Cambridge University Press, 2016).

14 Amartya Sen, 'Universal healthcare: the affordable dream,' *Guardian*, 2015.

15 Cereseto and Waitzkin, 'Economic development, political-economic system, and the physical quality of life,' 1986; Amartya Sen, 'Public Action and the Quality of Life in Developing Countries,' 1981; Vicente Navarro, 'Has socialism failed? An analysis of health indicators under capitalism and socialism,' *Science & Society*, 1993.

16 Jason Hickel, 'Is it possible to achieve a good life for all within planetary boundaries?' *Third World Quarterly* 40(1), 2019, pp. 18–35 (This research builds on Daniel O'Neill et al., 'A good life for all within planetary boundaries,' *Nature Sustainability*, 2018, p. 88–95); Jason Hickel, 'The Sustainable Development Index: measuring the ecological efficiency of human development in the Anthropocene,' *Ecological Economics* 167, 2020.

17 Ida Kubiszewski et al., 'Beyond GDP: Measuring and achieving global genuine progress,' *Ecological Economics* 93, 2013, pp. 57–68. The authors draw on Max-Neef to interpret this threshold as the point at which the social and environmental costs of GDP growth become significant enough to cancel out consumption-related gains. See Manfred

Max-Neef, 'Economic growth and quality of life: a threshold hypothesis,' *Ecological Economics* 15(2), 1995, pp. 115–118. See also William Lamb et al., 'Transitions in pathways of human development and carbon emissions,' *Environmental Research Letters* 9(1), 2014; Angus Deaton, 'Income, health, and well-being around the world: Evidence from the Gallup World Poll,' *Journal of Economic Perspectives* 22(2), 2008, pp. 53–72; Ronald Inglehart, *Modernization and Postmodernization: Cultural, Economic, and Political Change in 43 Societies* (Princeton University Press, 1997).

18 Tim Jackson, 'The post-growth challenge: secular stagnation, inequality and the limits to growth,' CUSP Working Paper No. 12 (Guildford: University of Surrey, 2018).

19 Mark Easton, 'Britain's happiness in decline,' BBC News, 2006.

20 Richard Wilkinson and Kate Pickett, *The Spirit Level: Why Equality is Better for Everyone* (Penguin, 2010).

21 Lukasz Walasek and Gordon Brown, 'Income inequality and status seeking: Searching for positional goods in unequal US states,' *Psychological Science*, 2015.

22 Adam Okulicz-Kozaryn, I. V. Holmes and Derek R. Avery, 'The subjective well-being political paradox: Happy welfare states and unhappy liberals,' *Journal of Applied Psychology* 99(6), 2014; Benjamin Radcliff, *The Political Economy of Human Happiness: How Voters' Choices Determine the Quality of Life* (Cambridge University Press, 2013).

23 According to the UN's World Happiness Report.

24 Dacher Keltner, *Born to be Good: The Science of a Meaningful Life* (W W Norton & Company, 2009); Emily Smith and Emily Esfahani, *The Power of Meaning: Finding Fulfilment in a World Obsessed with Happiness* (Broadway Books, 2017).

25 Sixty-year-old Nicoyan men have a median lifetime of 84.3 years (a three-year advantage over Japanese men), while women have a median lifetime of 85.1. See Luis Rosero-Bixby et al., 'The Nicoya region of Costa Rica: a high longevity island for elderly males,' *Vienna Yearbook of Population Research*, 11, 2013; Jo Marchant, 'Poorest Costa Ricans live longest,' *Nature News*, 2013; Luis Rosero-Bixby and William H. Dow, 'Predicting

mortality with biomarkers: a population-based prospective cohort study for elderly Costa Ricans,' *Population Health Metrics* 10(1), 2012.

26 Danny Dorling, *The Equality Effect* (New Internationalist, 2018).

27 Wilkinson and Pickett, *The Spirit Level*.

28 *Confronting Carbon Inequality*, Oxfam, 2020.

29 Yannick Oswald, Anne Owen, and Julia K. Steinberger, 'Large inequality in international and intranational energy footprints between income groups and across consumption categories,' *Nature Energy* 5(3), 2020, pp. 231–239.

30 Thomas Piketty, 'The illusion of centrist ecology,' *Le Monde,* 2019.

31 World Happiness Report.

32 CFO Journal, 'Cost of health insurance provided by US employers keeps rising,' *Wall Street Journal,* 2017.

33 David Ruccio, 'The cost of higher education in the USA,' *Real-World Economics Review* blog, 2017.

34 Average real wages peaked in 1973 at $23 per hour, declined to a nadir of $19 per hour in 1995, and stood at $22 per hour in 2018 (US Bureau of Labour Statistics). The poverty rate was 11% in 1973 and 12.3% in 2017 (US Census Bureau).

35 World Inequality Database.

36 See www.goodlife.leeds.ac.uk/countries.

37 Hickel, 'Is it possible to achieve a good life for all?' This research builds on Kate Raworth, "A safe and just space for humanity: can we live within the doughnut?" *Oxfam Policy and Practice* 8(1), 2012. Costa Rica is one of the strongest performers in this dataset, but it has relatively high levels of income inequality. This means it could improve social outcomes even further, without any additional growth. Note that the income poverty indicators in these studies rely on the World Bank's money-metric approach, which does not adequately account for changes in the prices of essential goods.

38 Joel Millward-Hopkins et al., 'Providing decent living with minimum energy,' 2020.

39 Frantz Fanon, The *Wretched of the Earth* (Grove Press, 1963).

40 See Ashish Kothari et al., *Pluriverse: A Post-Development Dictionary* (Columbia University Press, 2019).

41 Dorninger et al., 'Global patterns of ecologically unequal exchange.'

42 David Woodward, 'Incrementum ad absurdum: global growth, inequality and poverty eradication in a carbon-constrained world,' *World Economic Review* 4, 2015, pp. 43–62.

43 The 3 cents figure is based on World Bank poverty data from PovcalNet, excluding East Asia.

44 World Inequality Database.

45 According to World Bank data, the poverty gap at $7.40/day is $6 trillion, and the additional money needed to increase per capita health spending for low- and middle-income countries to the level of Costa Rica's is $4 trillion.

46 I draw this figure from the Credit Suisse Global Wealth Report, 2019.

47 Zak Cope, *The Wealth of (Some) Nations: Imperialism and the Mechanics of Value Transfer* (Pluto Press, 2019).

48 This figure comes from various reports by Global Financial Integrity.

49 This figure is based on estimates in the 1999 UN Trade and Development Report. The report indicates that $700 billion in potential revenues is lost each year in the industrial export sector, and more than this amount in the agricultural export sector.

50 I gleaned this insight from Dan O'Neill. See for instance: Rob Dietz and Daniel W. O'Neill, *Enough is Enough: Building a Sustainable Economy in a World of Finite Resources* (Routledge, 2013).

51 Data on global fossil fuel subsidies is from the IMF, and data on global military expenditure is from the World Bank.

52 Mariana Mazzucato, 'The entrepreneurial state,' *Soundings* 49, 2011, pp. 131–142.

Five: Pathways to a Post-Capitalist World

1 International Resource Panel, *Global Resources Outlook* (United Nations Environment Programme, 2019).

2 Bringezu, 'Possible target corridor for sustainable use of global material resources.' Nations need to get down to at most 8 tons of material footprint per person (Bringezu suggests a target of 3–6 tons per person by 2050). That means the US needs to reduce material use by 75%, the UK by

66%, Portugal by 55%, Saudi Arabia by 33%, etc., according to data for 2013 from materialflows.net. For more on the scale of necessary reductions across a variety of impact indicators see Hickel, 'Is it possible to achieve a good life for all?'

3 Joel Millward-Hopkins et al., 'Providing decent living with minimum energy'; Michael Lettenmeier et al., 'Eight tons of material footprint'.

4 Markus Krajewski, 'The Great Lightbulb Conspiracy,' *IEEE Spectrum*, 2014.

5 Data on appliance lifespans comes from 'How long should it last?' Whitegoods Trade Association. The WTA says that 'the average life-span has dropped from over ten years to under seven years and it is not unusual for cheaper appliances to only last a few years'. In the 'Study of Life Expectancy of Home Components', the National Association of Home Builders indicates that without planned obsolescence major appliances can last two to five times longer.

6 Data on global smartphone sales and global smartphone penetration comes from statista.com.

7 Alain Gras, 'Internet demande de la sueur,' *La Decroissance*, 2006.

8 Andre Gorz, *Capitalism, Socialism, Ecology*, trans. Chris Turner (London: Verso, 1994).

9 Robert Brulle and Lindsay Young, 'Advertising, individual consumption levels, and the natural environment, 1900–2000,' *Sociological Inquiry* 77(4), 2007, pp. 522–542.

10 Data on global advertising expenditures comes from statista.com.

11 Elizabeth Cline, 'Where does discarded clothing go?' *The Atlantic*, 2014.

12 Data from twenty-seven European countries from 1980 to 2011 shows an inverse relationship between advertising spending and citizens' sense of happiness and satisfaction. Nicole Torres, 'Advertising makes us unhappy,' *Harvard Business Review*, 2020.

13 I gleaned this insight from Noam Chomsky, from a 2013 interview conducted by Michael S. Wilson.

14 *Global Food: Waste Not, Want Not*, Institute of Mechanical Engineers, 2013.

15 These calculations simply assume halving total agricultural emissions (26% of global total) and land use (4.9 billion hectares). 'Food is responsible

for one-quarter of the world's greenhouse gas emissions,' Our World in Data, 2019; 'Land use,' Our World in Data, 2019.

16 'Grade A Choice?' Union of Concerned Scientists, 2012.

17 I say 'in most cases' because while the vast majority of beef is consumed as a commodity, there are some Indigenous or traditional pastoral communities (such as the Maasai in Kenya) that rely on cattle for subsistence.

18 Elke Stehfest et al., 'Climate benefits of changing diet,' *Climatic Change* 95(1–2), 2009, pp. 83–102.

19 Joseph Poore and Thomas Nemecek, 'Reducing food's environmental impacts through producers and consumers,' *Science* 360(6392), 2018, pp. 987–992.

20 Marco Springmann et al., 'Health-motivated taxes on red and processed meat: A modelling study on optimal tax levels and associated health impacts,' *PloS One* 13(11), 2018.

21 In the US, house sizes have grown from 551 square feet per person in 1973 to 1,058 square feet per person in 2015, US Census Bureau.

22 Fridolin Krausmann et al., 'Global socioeconomic material stocks rise 23-fold over the 20th century and require half of annual resource use,' *Proceedings of the National Academy of Sciences* 114(8), 2017, pp. 1880–1885.

23 Bringezu, 'Possible target corridor for sustainable use of global material resources.'

24 Multiple polls in the United States indicate strong majority support for a federal job guarantee. In the UK, it's 72% (YouGov, 2020).

25 For more on how a job guarantee could work, and how to fund it, see Pavlina Tcherneva, *The Case for a Job Guarantee* (Polity, 2020).

26 This research is reported in Kyle Knight, Eugene Rosa and Juliet Schor, 'Could working less reduce pressures on the environment? A cross-national panel analysis of OECD countries, 1970–2007,' *Global Environmental Change* 23(4), 2013, p. 691–700. It's interesting to note that the extra happiness accruing from free time is not positional, in contrast to happiness associated with income, so its benefits are durable. This same article reports on studies showing that people who work shorter hours have higher levels of well-being than those who work longer hours.

27 Anders Hayden, 'France's 35-hour week: Attack on business? Win-win reform? Or betrayal of disadvantaged workers?' *Politics & Society* 34(4), 2006, pp. 503–542.

28 This research is reported in Peter Barck-Holst et al., 'Reduced working hours and stress in the Swedish social services: A longitudinal study,' *International Social Work* 60(4), 2017, pp. 897–913.

29 Boris Baltes, et al., 'Flexible and compressed workweek schedules: A meta-analysis of their effects on work-related criteria,' *Journal of Applied Psychology* 84(4), 1999.

30 Anna Coote et al., '21 hours: why a shorter working week can help us all flourish in the 21st century,' New Economics Foundation, 2009.

31 François-Xavier Devetter and Sandrine Rousseau, 'Working hours and sustainable development,' *Review of Social Economy* 69(3), 2011, pp. 333–355.

32 See for example what happened in France when it shifted to a thirty-five-hour week: Samy Sanches, 'Sustainable consumption à la française? Conventional, innovative, and alternative approaches to sustainability and consumption in France,' *Sustainability: Science, Practice and Policy* 1(1), 2005, pp. 43–57.

33 David Rosnick and Mark Weisbrot, 'Are shorter work hours good for the environment? A comparison of US and European energy consumption,' *International Journal of Health Services* 37(3), 2007, pp. 405–417.

34 Jared B. Fitzgerald, Juliet B. Schor and Andrew K. Jorgenson, 'Working hours and carbon dioxide emissions in the United States, 2007–2013,' *Social Forces* 96(4), 2018, pp. 1851–1874.

35 This idea was articulated by Theodor Adorno and Max Horkheimer in *Dialectic of Enlightenment* (New York: Herder and Herder, 1972).

36 Lawrence Mishel and Jessica Schieder, 'CEO compensation surged in 2017,' Economic Policy Institute, 2018.

37 Sam Pizzigati, *The Case for a Maximum Wage* (Polity, 2018).

38 Pizzigati, *The Case for a Maximum Wage*.

39 World Inequality Database.

40 YouGov, 2020.

41 'Social prosperity for the future: A proposal for Universal Basic Services,' UCL Institute for Global Prosperity, 2017.

42 Frank Adloff describes this as an 'infrastructure of conviviality'. See his article 'Degrowth meets convivialism', in *Resilience*.

43 Walasek and Brown, 'Income inequality and status seeking: Searching for positional goods in unequal US states'.

44 And opportunities to learn and develop new skills such as music, maintenance, growing food and crafting furniture would contribute to local self-sufficiency. Samuel Alexander and Brendan Gleeson show how this works in their book *Degrowth in the Suburbs: A Radical Urban Imaginary* (Springer, 2018).

45 Kallis, *Limits*, p. 66.

46 There is evidence of this from studies done in Canada, Italy and the UK, reported in Stratford, 'The threat of rent extraction'.

47 Graeber, *Debt*.

48 Graeber, *Debt*, p. 82.

49 Johnna Montgomerie, *Should We Abolish Household Debts?* (John Wiley & Sons, 2019).

50 I discuss this history in detail in *The Divide*.

51 Some cities and regional governments have experimented with 'citizens' debt audits' whereby people decide collectively which debts can be cancelled without social fallout, and which should be repaid. In order to prevent a lending crisis, cancellations should be conducted in a phased manner, and a parallel public banking system should be established that is ready to lend and keep up confidence even if over-exposed banks go under.

52 Graeber, *Debt*, p. 390.

53 Thanks to Charles Eisenstein for this analogy.

54 Louison Cahen-Fourot and Marc Lavoie, 'Ecological monetary economics: A post-Keynesian critique,' *Ecological Economics* 126, 2016, pp. 163–168.

55 Mary Mellor, *The Future of Money* (Pluto Press, 2010).

56 *Escaping Growth Dependency* (Positive Money, 2020); Stephanie Kelton, *The Deficit Myth: Modern Monetary Theory and How to Build a Better Economy* (Hachette UK, 2020); Jason Hickel, 'Degrowth and MMT: A thought experiment', 2020 (www.jasonhickel.org/blog/2020/9/10/degrowth -and-mmt-a-thought-experiment).

57 Oliver Hauser et al., 'Co-operating with the future,' *Nature* 511(7508), 2014, pp. 220–223.

58 This data on lobbying comes from the Centre for Responsive Politics.

59 Raquel Alexander, Stephen W. Mazza, and Susan Scholz, 'Measuring rates of return on lobbying expenditures: An empirical case study of tax breaks for multinational corporations,' *Journal of Law & Politics* 25, 2009.

60 Martin Gilens and Benjamin I. Page, 'Testing theories of American politics: Elites, interest groups, and average citizens,' *Perspectives on politics* 12(3), 2014, pp. 564–581.

61 Simon Radford, Andrew Mell, and Seth Alexander Thevoz, '"Lordy Me!" Can donations buy you a British peerage? A study in the link between party political funding and peerage nominations, 2005–2014,' *British Politics*, 2019, pp. 1–25.

62 Ewan McGaughey, 'Democracy in America at work: the history of labor's vote in corporate governance,' *Seattle University Law Review* 697, 2019.

63 'Media Ownership Reform: A Case for Action,' Media Reform Coalition, 2014.

64 Ashley Lutz, 'These six corporations control 90% of the media in America,' *Business Insider*, 2012.

65 Elinor Ostrom, *Governing the Commons: The Evolution of Institutions for Collective Action* (Cambridge University Press, 1990).

66 I gleaned this idea from the Greek-French philosopher Cornelius Castoriadis.

Six: Everything is Connected

1 Interviewed by ethnographer Knud Rasmussen in the early twentieth century.

2 Lourens Poorter et al., 'Biomass resilience of Neotropical secondary forests,' *Nature* 530(7589), 2016, pp. 211–214.

3 Susan Letcher and Robin Chazdon, 'Rapid recovery of biomass, species richness, and species composition in a forest chronosequence in northeastern Costa Rica,' *Biotropica* 41(5), pp. 608–617.

4 In what follows I draw on ethnographic material discussed by Philippe Descola in *Beyond Nature and Culture* (University of Chicago Press, 2013).

5 Graham Harvey, *The Handbook of Contemporary Animism* (Routledge, 2014).

6 Following Graham Harvey, I'm referring here to Martin Buber's distinction between I-thou and I-it paradigms.

7 For this I draw on the work of Eduardo Viveiros de Castro and his notion of 'perspectivism'. See, for instance, 'Cosmological deixis and Amerindian perspectivism,' *Journal of the Royal Anthropological Institute*, 1998.

8 Hannah Rundle, 'Indigenous knowledge can help solve the biodiversity crisis,' *Scientific American*, 2019.

9 For more on Spinoza's naturalism, see Hasana Sharp, *Spinoza and the Politics of Renaturalization* (University of Chicago Press, 2011).

10 David Abram, *The Spell of the Sensuous: Perception and Language in a More-Than-Human World* (Vintage, 2012).

11 This research is reported by Carl Zimmer, 'Germs in your gut are talking to your brain. Scientists want to know what they're saying,' *New York Times*, 2019.

12 Jane Foster and Karen-Anne McVey Neufeld, 'Gut–brain axis: how the microbiome influences anxiety and depression,' *Trends in Neurosciences* 36(5), 2013, pp. 305–312

13 John Dupré and Stephan Guttinger, 'Viruses as living processes,' *Studies in History and Philosophy of Science Part C: Studies in History and Philosophy of Biological and Biomedical Sciences* 59, 2016, p. 109–116.

14 Ron Sender, Shai Fuchs and Ron Milo, 'Revised estimates for the number of human and bacteria cells in the body,' *PLoS Biology* 14(8).

15 John Dupré, 'Metaphysics of metamorphosis,' *Aeon*, 2017.

16 Robert Macfarlane, 'Secrets of the wood wide web,' *New Yorker*, 2016.

17 Brandon Keim, 'Never underestimate the intelligence of trees,' *Nautilus*, 2019. On plant learning and memory, see Sarah Lasko, 'The hidden memories of plants,' *Atlas Obscura*, 2017.

18 Andrea Morris, 'A mind without a brain. The science of plant intelligence takes root,' *Forbes*, 2018.

19 Josh Gabbatiss, 'Plants can see, hear and smell – and respond,' *BBC Earth*, 2017.

20 Keim, 'Never underestimate the intelligence of trees.'

21 Chorong Song et al., 'Psychological benefits of walking through forest areas,' *International Journal of Environmental Research and Public Health* 15(12), 2018.

22 Jill Suttie, 'Why trees can make you happier,' *Thrive Global*, 2019. I credit Suttie's work with pointing me to many of the studies I mention here.

23 Ernest Bielinis et al., 'The effect of winter forest bathing on psychological relaxation of young Polish adults,' *Urban Forestry & Urban Greening* 29, 2018, pp. 276–283.

24 Geoffrey Donovan et al., 'Is tree loss associated with cardiovascular-disease risk in the Women's Health Initiative? A natural experiment,' *Health & Place* 36, 2015, pp. 1–7.

25 Bum-Jin Park et al., 'The physiological effects of Shinrin-yoku (taking in the forest atmosphere or forest bathing): evidence from field experiments in 24 forests across Japan,' *Environmental Health and Preventive Medicine* 15(1), 2010.

26 Bing Bing Jia et al., 'Health effect of forest bathing trip on elderly patients with chronic obstructive pulmonary disease,' *Biomedical and Environmental Sciences* 29(3), 2016, pp. 212–218.

27 Qing Li et al., 'Effect of phytoncide from trees on human natural killer cell function,' *International Journal of Immunopathology and Pharmacology* 22(4), 2009, pp. 951–959.

28 Omid Kardan et al., 'Neighbourhood greenspace and health in a large urban centre,' *Scientific Reports* 5, 2015.

29 Robin Wall Kimmerer, *Braiding Sweetgrass: Indigenous Wisdom, Scientific Knowledge and the Teachings of Plants* (Milkweed Editions, 2013).

30 Marcel Mauss' book *The Gift* has been fundamental to degrowth thinking.

31 Rattan Lal, 'Enhancing crop yields in the developing countries through restoration of the soil organic carbon pool in agricultural lands,' *Land Degradation & Development* 17(2), 2006, pp. 197–209.

Acknowledgements

1 I draw this language of 'trance', and the story of the desert crossing, from the work of Tara Brach.